Spices and Condiments

Origin, History and Applications

Spices and Condiments

Origin, History and Applications

By
Dr. D.A. Patil
Professor

Dr. D.A. Dhale
Assistant Professor

*P.G. Department of Botany,
S.S.V.P. Sansthans, L.K. Dr. P.R. Ghogrey Science College,
Dhule – 424 005, (M.S.) India*

2013
Daya Publishing House®
A Division of
Astral International Pvt. Ltd.
New Delhi – 110 002

Published by	:	**Daya Publishing House**®
		A Division of
		Astral International Pvt. Ltd.
		ISO 9001:2008 Certified Company
		–4760-61/23, Ansari Road, Darya Ganj
		New Delhi-110 002
		Ph. 011-43549197, 23278134
		E-mail: info@astralint.com
		Website: www.astralint.com
Laser Typesetting	:	**Classic Computer Services**
		Delhi - 110 035
Printed at	:	**Chawla Offset Printers**
		Delhi - 110 052

PRINTED IN INDIA

Preface

We, human beings, live on the earth planet as guest of green plants which provide not only the basic amenities of human life but also add pleasure and beauty to it. One such fascinating group of plant is the 'Spices and Condiments'. The great mystery and beauty of this plant group lies in their use, blending and ability to change and enhance the character of food. They are virtually indispensable in the culinary art. They have been used for culinary, flavoring and preservation purposes since the dawn of recorded history and undoubtedly even before. They have special significance in various ways in human life because of their flavor, taste and aroma. Recently, they have been reputed to cure many diseases and have also served as basis of perfumes, balms and incense. These fascinating useful plants not only treat their consumers, but also entice to write on them. Their history itself is self-explanatory in this regard. The story of spices and condiments is one of the most romantic chapters in the history of plant products as they are connected with many important events in man's history, including economic welfare, geographical discovery, annexation of territories and all the vices of theft, envy and hatred of which man is capable. Spices have been responsible even for the rise and fall of empires of the earth globe. Their search leads to the discovery of new continent and waterways. No wonder the present authors are also in their love, love to write.

This book is written although specially for the University and college students and teachers in Botany and Agriculture but will be also useful in the studies of ethnobotany, food sciences, archaeology, nutrition, medicine or indeed to anyone with a concern for natural resources on the Earth. Anecdotes and stories concern with spices and their history will certainly render this book more readable even to the layman. Book on spices and condiments may be too many but attempts to collate information from various domains of plant sciences and social sciences in a book of this kind are very few. This is hoped to increase general readership of the book.

This book covers the nomenclature, brief description, distribution, varieties if any, origin, history, ancient applications if any, of spices and condiments of the world. The useful parts, chemical constituents, properties and worldwide uses, medicinal uses and more importantly processed products normally with wide range of available information in the world market are given pertinently. All these aspects are presented together in the hope of rendering some humble service to those interested in studying spices and condiments. Readers in all walks of life are humbly requested to comment on the contents and subject matter of the book. Their response may help for further improvements.

Prof. D.A. Patil

Dr. D.A. Dhale

Contents

Part I

Introduction

Chapter 1
Introduction

Spices, since antiquity, have been considered indispensable in the culinary art for flavouring foods. Some are used in pharmaceuticals, perfumery, cosmetics and several other industries. Many of them possess colourant, preservative, antioxidant, antiseptic and antibiotic properties. The word 'spice' is derived from the Latin word 'species', meaning specific kind. The name implies a fact that all plant parts have been cultivated for their aromatic, fragrant, pungent or any other desirable properties including the seed (aniseed, caraway, coriander), leaf (cilantro, kari, bay, mint), berry (allspice, juniper, black pepper), bark (cinnamon), kernel (nutmeg), aril (mace), stem (chives), stalks (lemongrass), rhizome (ginger, turmeric, galangal), root (lovage, horseradish), flower (saffron), bulb (garlic, onion), fruit (star anise, cardamom, chile pepper) and flower bud (clove).

According to the latest available International Organization for Standardization (ISO) report, there is no clear-cut division between 'spice' and 'condiments' and as such they have been clubbed together. The terms 'spice' and 'condiments' applies to such natural plants or vegetable products or mixtures thereof, used in whole or ground form, mainly for imparting flavour, aroma and piquancy to foods and also for seasoning of food and beverages like soups, etc.

Chapter 2
In Search of Spices

In the medieval times India in the mind of foreigners was a land of 'maharajas', 'diamonds', 'fine textiles', 'ivory' and, of course, 'spices' too. The world still visaulise India as the real 'Home of Spices'. This is so because of quality of spices produced and exported from India, has been and continued to be undisputedly the best. In ancient times, spices ranked with precious stones in the inventory of royal possessions. Spices determined the wealth and policies of nations and also played an important role in ancient medicine.

Spices have been eagerly sought by man long before the recorded history. Spices, in ancient times, were so extraordinarily expensive that only the wealthy people could afford them. Spices with strong and pungent flavours have been used in the past not only in the cookery, but also for preserving food before the advent of refrigeration. Aroma and flavour have been a part of the magical rites of the Shaman, and have also been employed for various 'purification' ceremonies. The streets were often fumigated with spices during the visit of royal guests. Ancient Egyptians used them to perfume the person and to embalm the dead, the bodies of the deceased being filled with the purest myrrh, cassia and other spices.

Even before the birth of Greece and Rome, the sailing ships carried Indian perfumes, spices and textiles to Mesopotamia, Arabia and Egypt. It was the lure of these commodities that brought many sailors to the shores of India.Long before the Christian era, the Greek merchants crowded the market of South India for buying spices and other precious things. Eupicurean Rome was spending a fortune on

Indian spices, silk, brocades and cloth of gold. The Parthian was believed to have been fought by Rome. All these were mainly aimed at keeping open the trade route to India.

The quest of the spices of the early European explorers led to the discovery of new world and water ways. The westward voyages of Christopher Columbus were basically aimed at searching out the 'Spice Island' of the Far East. Although he failed to reach the destiny, it resuled in the discovery of America. Unintentionally Columbus and his men discovered some valuable and entirely new species, such as allspice and red pepper in the New World.

The search of spices and condiments is the most romantic and fascinating history. Search of spices was connected with many important events including geographical discovery, economic warfare, annexation of territories and all the vices of theft, envy and hatred of which man is capable. Spices have been basic cause for the rise and fall of empires and the great sea voyages to explore the distant corners of the globe. Da Gama's triumphant voyage intensified an international power struggle over spices. For three hundred years later the Western countries *e.g.,* Portugal, Spain, France, Holland, and Great Britain fought wars for spice-producing colonies and the control of the spice trade. During the middle ages, a pound of ginger was worth a sheep, a pound of mace worth three sheeps or half a cow. Pepper was the most valuable spice of all. It was counted out in the individual pepper-corns and a sack of pepper was said to be worth a man's life. Then in 1605, the Dutch, however, could not maintain their monopply of spice trade for long. Even their extremely stern and repressive measures to control the monopoly failed. The Dutch were thwarted in their efforts not only by the French, but by pigeons that swallowed nutmeg seeds and carried them to nearby Islands. Later, India and nearby spice Islands were under British regime. They established the East India Company and shared with Holland most of the spice trade. In 1672, Elihu Yale joined the British East India company as a clerk. The fortune he made in the spice trade resulted in the foundation of Yale University in the United States.

During early part of the 18th century, spices were smuggled away and planted around the world, especially in the West Indies, Zanzibar, Madagascar, Malaysia and Srilanka. This ended scarcity and monopoly. Nowadays, substantial plantations are grown in

America. The best quality cardamoms now hail from Guatemala, the finest nutmeg and mace from Grenada and select black pepper from Brazil. However, the vast majority of spices are still obtained from Asia. The African country gave to mankind grains of paradise, *Aframomum melegueta* (Rosc.) Schum. (Zingiberaceae) and Madagascar clove, *Ravensara aromatica* Gmel. (Lauraceae), while American tropics provided vanilla, the capsicum and allspice. The colder region of northern Europe and Asia produced caraway, coriander, cumin and mustard seed.

The cultivation and use of spice plants goes back to the beginning of history and most of the spices were known to the ancient civilization of China, Egypt, Greece, Rome, the East Indies and rest of the Old World since the early days of history. Spices and herbs used to play an important part in the life of these countries. Spices were among the first items to be traded between the East and the West. The Persians were the first traders who used to transport the product of India and the neighbouring Molucca islands by camel caravans. They sold them to the Phoenicians who traded them all along the Mediterranean coast from Alexandria to Rome. The Indian and Greek traders monopolized the trades in spice later from the first to the 8th century A.D. The Arabs controlled the trade in spices up to the 14th century. They sold spices to Egyptian, Greek and Venetian traders. Spices gained increasing popularity in the West for uses such as preservative and flavouring food on ships during long voyages and for seasoning favourite dishes. It was a lucrative business. The Venetians misconceived that spices were produced somewhere in Arabia. The Arabs intentionally concealed the source of the spices and their processing. They spread false rumours in procuring spices.

Marco Polo's 24-year journey that began in 1271 from Venice and took him through the Far East. This brave attempt solved the mystery of the origin of the Arabs' supply of spices. The high duties imposed by the rulers of Egypt on spices were partly responsible for European voyagers seeking another route to India. The magic of the spice trade lured adventures from the Western world to the Indian ocean islands. Spice trade brought fame to many of them such as Diaz, Da Gama, Columbus, Magellan, Drake, Cavendish and Van Houtman. Vasco da Gama succeeded in rounding up the Cape of Good Hope and reaching India by sea in 1498. Only two of Da

Gama's four ships survived to reach their homeport after journey of 24000 miles in two years. These two surviving ships brought back a cargo of spices and other products worth sixty times the cost of the voyage. Following the circumnavigation of Africa, the Portuguese gained control of much of the Indian ocean and extended their trade as far as China. Goa, in India, remained a Portuguese possession until recently. They virtually monopolized the spice trade for well over a century.

Chapter 3
Forms and Composition of Spices

Spices are available in different forms: fresh, dried, whole, ground, crushed, pureed, as extractives and pastes. Each form has its own qualities and drawbacks. These forms depend on the specific application, processing parameters and shelf-life.

Fresh Whole Spices

The taste of fresh food is always preferred. Consumers seek fresh taste from spices initially from their aroma. This aroma is because of the volatile component of the spice. It can be lost during harvesting, storing, processing or handling. In case of certain spices, fresh forms have different flavor profiles than dried forms, *e.g.*, ginger, cilantro or basil. Fresh ginger has been found to be less pungent than dried ginger since fresh ginger has less shogaols (non-volatile constituents that cause pungency) than dry ginger.

Whole spices when freshly ground give prepared food a fresh taste. Fresh spices provide crunchy, crispy textures and colourful appeal. Fresh whole spices also become very aromatic when they are roasted or fried in oil, and their aroma transfer to the application. This is especially a case of whole or cracked seeds *e.g.*, mustard seeds, bay leaf, kari leaf. Whole spices provide aroma, texture and visual effect. Certain spices have a strong aroma when fresh, *e.g.*, basil, garlic, onion and cilantro, on account of their high volatile (essential) oils. The essential oils fuse quickly at high temperatures, especially if the spices are processed in aqueous system. They can also be lost at room temperatures or even when the spices are cut or bruised.

Flavours, aroma, colors, textures are divulged by certain applications in various ways: (a) Uneven distribustion of whole spices in a product is sometimes to achieve nuttiness or a sensation of 'bite' into a whole spice, *e.g.* whole sesame seeds on a breadstick or ajowan seeds on Indian naan bread. In such cases, whole spices can become the major flavor characterizing a product. As whole spices, especially the leafy spices, provide a great visual appeal as garnishes. (b) Flavor is intact in the whole spice. It is more slowly released than with the ground spice, especially when subjected to preparation techniques like frying or roasting, during which time, the whole spice slightly cracks open. (c) In whole spice, the chemical compounds that provide the flavors vary in concentration throughout the spice. In chile peppers, the greatest concentrations of pungent compounds are found in the inner portions, such as the veins and seeds. (d) In many whole spices, cooking or processing changes chemical compounds and their proportions to be varing degrees, often giving rise to different flavors. For example, smoking, grilling or drying certain chile peppers significantly changes their flavor and color completely, giving it a new identity, called chipotle. (e) Spices lacking a strong aroma, *e.g.* bay leaf, chile pepper or sesame seed, develop intense flavor after roasting or boiling. Mustard seed, star anise and fagara (Szechwan pepper) are generally dry roasted to intensify their flavors for meat, fish and poultry dishes. (f) Certain spices, such as lemongrass, spearmint, basil and chile peppers, are blended fresh and are used in making sauces and condiments with water, oil, wine or vinegar. The fresh pureed or paste forms have intense flavors and need to be mixed well before application in sauces, soups or gravies. Since the paste form usually contains oil, it can become rancid in a shorter period of time. (g) Consumers want to use 'fresh' spices, but usually their flavors, colors and textures are lost during storage and prolonged processing. Preliminary preparations, *e.g.* grinding, roasting or flaking of whole spices, need to be done before adding them to processed food. (h) Consistency is difficult to achieve in fresh spices as their origin, age and storage conditions cause flavor variations. Therefore, dry spices and spice extractives are, by necessity, the forms most often used to formulate foods or beverages. Fresh whole spices are not frequently found in processed foods, but are generally used in restaurants, in home cooking and in other smaller scale applications. (i) The goal for a food designer is to develop products that will have the 'fresh' quality desired by consumers but that have spice-sensory attributes that can withsand processing, freezing and storage conditions.

Dried Spices

Spices are often used in dried forms since (a) they are not subject to seasonal availability, (b) are easier to process, (c) have longer shelflife and (d) have the lowest cost. They are used for processed products or for wholesale usage. They are available whole, finely or coarsely ground, cracked and in various sized particulates. Grinding generates rapid air movement and heat that dissipate some of the volatile oils and also change some natural flavor notes through oxidation.

The same dried spice delivers different flavor perceptions in finished product. Ground spices have better dispersibility in food products than fresh whole spices. Volatile oils are released through grinding since it partially breaks down the cellular matrix of the spice. Flavor is intensified through drying because of the elimination of most moisture and concentration of the low volatile compounds that give stronger flavor However, this renders it less aromatic due to the loss of the volatile constituents. Dried spices can better withstand the higher temperature and processing conditions than fresh spices.

Some dried spices characterize flavor and texture of an application. For example, garlic and onion, which are powdered, granulated, ground, minced, chopped, and sliced and in various sized particles, characterize flavor and texture in garlic bread, onion bagels or chops. Some ground spices need to be 'rehydrated' in order to develop their flavor, for example, ground mustard that becomes pungent only when water is added. This addition of water triggers an enzyme reaction that releases aroma of the spice. Acidulates, oil or vinegar are also added to preserve the pungency or intense flavors of spice in the finished product.

In case of processed foods, dried spices can be more economical to use than fresh spices. For example, dried leafy spices do not require the cutting, chopping or grinding preparation that the fresh forms do. Moreover, most dried spices retain a higher overall flavor concentration than fresh spices. e.g, one pound of dried garlic has equivalent flavor of five pounds of fresh garlic.

The same type of spice can have different sensory characteristics depending on where it was grown and how it was harvested, stored and processed. For example, dried ginger from India has a subtle lemon like flavor, dried ginger from southern China comes with slightly bitter notes and ginger from Jamaica has more pungent flavor. Similarly, ground black pepper which hails from a dried berry called peppercorns from Tellicherry are highly aromatic (India), while Lampong (Sumatra) pepper is less aromatic with more pungency. The Malaysian and Brazilian peppercorns, in contrast, have middle aromas with stronger bites.

The time period between harvesting and storage, and between grinding and adding to a food are crucial for obtaining its maximum potential. Flavor and color depend on the way a spice is treated or processed before being ground, and the conditions of storage before delivery to the food processor. Spice flavor can be readily oxidized, and losses occur during milling and storage of spices.

Some spices *e.g.* cumin, coriander and cardamom posses more aroma and flavor when freshly ground than when bought as a pre-ground spice. On the grinding the spices, the oils tend to volatilize, causing aroma losses. Anise, black pepper and allspice lose their aroma quickly as soon as they are ground. To retain better color, flavor and aroma, spices are sometimes milled using lower temperatures. The finely ground spices blend better in finished products that require a smooth texture.

Ground spices should not be exposed to light, high temperatures or humidity and should be stored in tightly closed containers. Moisture and high temperatures will help mold growth that will cause spoilage. Generally, the moisture content of spice is 8-10 per cent. High storage temperature cause flavor loss, colour changes and caking or hardening of the ground spice. Spice need to be stored at 10-15°C with a relative humidity (RH) of 55-65 per cent.

Some dried spices have poor flavor intensity, can cause discoloring and create an undesirable appearance in the product. For example, dried ground cayenne can cause irregular variations in flavor and color, sometimes creating 'hot' spots in food products. Anticaking agents are added to ensure better flowability of dried spices. In applications with high moisture content, *e.g.* salad dressings or soups, where particulates are desired for visual and textural effects, there is a great risk in using dried spices, unless they are sterilized.

Spice Extractives

The sensation like sweet, piney, sour, bitter, spicy, sulfury, earthy and pungent are derived from an overall combination of aroma (due to volatile components) and taste (mainly due to nonvolatile) components in a spice. Spice extractives, which are highly concentrated forms of spices, contain the volatile and non-volatile oils that give each spice characteristic flavor. The volatile portion of spice extractives, also referred to as essential oils, typify the particular aroma of the spice. Most of the spices owe their distinctive 'fresh' characters to their essential oil content. The non-volatiles include fixed oils, gum resins, antioxidants and hydrophilic compounds, and they contribute to the taste or 'bite' of spice. Certain spices are prized for their bites and coloring, such as black pepper, chile, peppers, ginger, saffron and turmeric. These properties are due to non-volatile portions of spices. Several chemical compounds are found in volatile oil whose amount and proportion give rise to the spices characteristic aromas. These can include one, two or several components. The major chemical components of essential oils are terpene compounds (monoterpenes, diterpenese and sesquiterpenes). Monoterpenes are the most volatile of these terpenes and give out strong aroma when tissues and cells of spices are disintegrated through heating, crushing, slicing or cutting.

The taste of spice like sweet, spicy, sour or salty, is due to esters, phenols, acids, alcohols, alkaloids and sugars. Sweetness is due to esters and sugars; sourness due to organic acids (citric, malic, acetic or lactic); saltiness due to cations, chlorides and citrates; astringency due to phenols and tannins; bitterness due to alkaloids (caffeine and glycosides), and pungency due to the acid-amides, carbonyls, thioethers and iso-thiocyanates.

The ratio of volatiles to non-volatiles varies among spices causing flavor similarities or differences within a genus and even within a variety. Within the genus *Allium* for example, there are differences in flavor among garlic, onions, chives, shallots and leeks, which differ in this ratio. They vary depending upon the species of spice, its source, environmental growing, harvesting conditions, storage and preparation methods. The distillation techniques can also give rise to varying components through loss of high boiling volatiles, with some components not being extracted or with some undergoing changes. Certain non-volatiles in a spice also differ with variety, origins, environmental growth conditions, stage of maturity and post-harvest conditions. For example, the different chile peppers belonging to the *capsicum* group, such as ha-baneros, cayennes, jalapenos or poblanos, all have distinct flavor perceptions, depending on the proportion of the different non-volitiles, the capsaicinoids.

Some spice extractives are available as natural liquids (which include essential oils, oleoresins and aquaresins) and dry encapsuled oils (spray dried powders and dry solubles). Developed from fresh or coarsely ground spices, spice extractives are standardized for their antioxidant activity. They are better concentrated than dried or fresh spices and so are used at much lower levels. These extractives provide better consistency than dried spices in prepared foods.

Essential (Volatile) Oils

Essential oils are the principal flavoring constituents of spice. Each essential oil has many chemical cpmponents, sometimes even up to 15, but the characterizing aroma generally constitutes anywhere from 60-80 per cent of the total oil. The essential oils are composed of hydrocarbons or terpenes (α-terpinene, α-pinene, camphene, limonene, phellandrene, myrcene and sabinene), oxygenated derivatives of hydrocarbons (linalool, citronellol, geraniol, carveol, menthol, borneol, fenchone, tumerone and nerol), benzene compounds (alcohols, acids, phenols, ester, and lactones) and nitrogen or sulfur containing compounds (indole, hydrogen sulfide, methyl propyl disulfide and sinapine hydrogen sulfate). Terpenes usually contribute to the freshness of a spice and can be termed floral, earthy, piney, sweet or spicy. The oxygenated derivatives, which include alcohols, esters, acids, aldehydes and ketones, are the major contributors to the aromatic sensation of a spice. The compounds with benzene structure provide sweet, creamy and floral notes, while the sulfur and nitrogen containing compounds afford the characteristic notes to onion, garlic, mustard, citrus and floral oils.

Essential oils are soluble in alcohol or ether and are only slightly soluble in water. They provide more potent aromatic effect than the

Table 1: Essential Oil Components of Spices

Sl.No.	Spice	Components in Essential Oils
1.	Allspice	Eugenol; 1,8-cineol; humulene, α-phellandrene
2.	Basil	Linalool ; 1,8-cineol ; methyl chavicol, eugenol
3.	Cardamom	1,8-cineol, linalool; limonene, α-terpineol acetate
4.	Dill leaf	Carvone, limonene, dihydrocarvone, α-phellandrene
5.	Epazote	Ascaridol, limonene, *p*-cymene, myrcene, α-pinene
6.	Fennel	Anethole, fenchone, limonene, α-phellandrene
7.	Ginger	Zingiberene, curcumene, farnescene, linalool, borneol
8.	Juniper	α-pinene, β-pinene, thujene, sabinene, borneol
9.	Kari leaf	Sabinene, α-pinene, β-caryophyllene
10.	Lemongrass	Citral, myrcene, geranyl acetate, linalool
11.	Marjoram	*cis*-sabinene, α-terpinene, terpinene 4-ol, linalool
12.	Nutmeg	Sabinene, α-pinene, limonene, 1,8-cineol
13.	Oregano	Terpinene 4-ol, α-terpinene, *cis*-sabinene
14.	Pepper, black	Sabinene, α-pinene, β-pinene, limonene, 1,8-cineol
15.	Rosemary	1,8-cineol, borneol, camhor, bornyl acetate
16.	Star anise	Anethole, α-pinene, β-phellandrene, limonene
17.	Turmeric	Turmerone, dihydrotumerone, sabinene, 1,8-cineol
18.	Zeodary	Germacrone-4, furanodienone, curzerenone, camphor.

ground spices. Essential oils lose their aroma with age. They are very concentrated, about 75-100 times more concentrated than the fresh spice. They do not have complete flavor profile of ground spices, but they are used where a strong aromatic effect is desired. Essential oils are used at very low level of 0.01-0.05 per cent in finished products.

Oleoresins

The non-volatile components create the heat or pungency of black pepper, mustard, ginger and chile peppers. These components can be acid amides, such as capsaicin in red pepper or piperine in black pepper, isothiocyanates in mustard, carbonyls such as gingerol in ginger and thioethers such as the di-allyl sulfides in garlic or onion. The different pungent or heat principles give different sensation *e.g.* spicy hot, hot, sharp, bitter or sulfury. The pungent sensation of onion or garlic is sulfury, while that of Jamaican ginger is spicy. Red pepper and white pepper do not contain much aroma because they are having very little essential oils, whereas ginger, black pepper and mustard contribute aromatic sentations with their bites because of higher content of volatile oils. White pepper posses a different bite sensation than black pepper because of their proportions of piperine and chavicine contents.

In chile peppers, five types of capsacinoids have been isolated: (i) capsaicin, (ii) hydrocapsaicin, (iii) dihydrocapsaicin (iv) dihydrohomocapsaicin and (v) homocapsaicin; each with its own characterizing 'bite' sensation in the mouth. In any particular type of chile pepper, the levels of capsaicinoids differ, causing varying heat levels. Each type of capsaicinoid also creates a different perception of heat. Habanero has an initial sharp and violent bite that quickly disappears, leaving behind an aromatic sensation, whereas the cayennes give an initial burn that lingers.

The release of heat sensation in mustard is different from wasabi. In wasabi, heat is immediate and in front of mouth, while with mustard and horseradish, the release is delayed and senses at the back of mouth, with a shooting sensation to the sinuses. Some of the non-volatiles that contribute pungency to a spice are given Table 2.

Table 2: Hot and/or Pungent Non-Volatile Spices

Sl.No.	Spice	Pungent Components of Spices
1.	Red pepper	Capsaicin, hydrocapsaicin, dihydrocapsaicin, dihydrohomocapsaicin, homocapsaicin.
2.	Black/white pepper	Piperine, chavicine
3.	Sansho pepper	Sanshool
4.	Mustard	Allyl isothiocyanate
5.	Horseradish	Allyl isothiocyanate
6.	Ginger	Gingerol, shogoal
7.	Garlic	Diallylsulfide
8.	Onion	Diallylsulfide

Oleoresins are available as viscous oils and thick pastes and are more difficult to handle than essential oils. Usually, oleoresins are mixed with a diluent such as propylene glycol, glycerol or other oils for better handling. An emulsifier is added for use in beverages, sauces, soups, pickles and salad dressings.

Being highly concentrated, oleoresins are used at very low concentrations. They have greater heat stability than essential oils. They give more uniform flavor and color with less variability than their ground spice counterparts. They are typically used in high heat applications such as soups, salad dressings, processed meats and in dry mixes and spice blends. Aquaresins are oleoresins that are dispersible in water and oil. They are convenient to use because of the ease with which they disperse into water-based foods such as soups, sauces, pickles or gravies.

Other Extractives

Liquid soluble spices are the blends of essential oils or oleoresins which are made for aqueous systems. Fat based soluble spice is made from essential oil or oleoresin blended with vegetable oil and used for mayonnaise, sauce or soup. Dry soluble spices, usually used in dry blends, are prepared by dispersing standardized extractives on a carrier such as salt, dextrose or maltodextrin. Encapsulated oils are prepared using essential oils and/or standardized oleoresins with gum arabic or modified as starch as the encapsulent. These have 5-10 times the strength of dried ground spices. Spray-dried flavors are the traditional encapsulated products. Spray-dried spice flavors or dry soluble spices are created to make liquid spices or extractives more convenient to handle and use in dry applications. They are a dispersion of up to 5 per cent or more of spice oleoresin on a free-flowing carrier such as salt, dextrose, gum Arabic, modified starch or maltodextrin. These encapsulated spice extractives are employed in high temperature applications, *e.g.* baked or retorted products. The spice flavors are slowly released into the product at the appropriate processing temperatures.

Oleoresins or essential oils are encapsulated to keep the full flavor impact of spices over the longer shelflife. This process grinds and encapsulates the spices in a closed system so that no volatile oils can escape. They are encapsulated by creating an emulsion with modified starch, dextrose and maltodextrin or soluble gum and spray dried under controlled temperature and humidity conditions. The spice extractives are entrapped in this matrix. This protects the flavor from oxidation and high heat and thereby provides longer shelf-life.

Encapsulated oleoresins retain the fresh notes of spices better than the oleoresins. They have no particulates and are completely natural, and like essential oils or oleoresins have a friendly ingredient label. They are water soluble and permit flavor to be liberated uniformly throughout the food. For application, a 1:1 or 1:2 ratio is used as a replacement for the non-capsulaled extractives. Encapsulation of an extractives makes it wettable and dispersible in water or oil and also decreases the dusting in production. The quantity of extractives that can be dispersed on the carrier depend with the type of carrier and the oleoresin. It is important to evenly disperse and blend the oleoresin onto the carrier.

Extractives of the spice are cost effective in comparison to fresh or dried spice as they are used at very low concentrations and provide similar or sometimes more acceptable flavor perceptions. One part oleoresin is equivalent to 20-40 parts of a ground spice. The color texture and flavor of dried and fresh spices are also altered through heat and freezing, while extractives have some heat and freezer stability. Extractives are available throughout the year and have standardized flavor and color, whereas the dried or fresh spices fluctuate in availability and quality. Extractives are also free of microbes and other extraneous matter, so they do not cause microbial contamination in the finished product.

Extractives are labeled as natural flavors, natural flavorings or as spice extracts. They are typically used by food developers because of their consistency in flavors and aroma, instant flavor release, uniformity of color and stability in high heat applications. The quality and consistency of products from development through production can be better controlled by using natural extractives.

Chapter 4
Functions of Spices

Spices have primary and secondary functions in food products. Their primary functions are to flavor food and provide aroma, texture and color. They also have secondary functions such as preservative, nutritional and health functions.Spices contain fibre, carbohydrates, fats, sugar, protein, gum ash, volatile (essential oil) and other non-volatile components. All of these components impart particular flavor, color, nutritional, health or preservative effect of each spice. The flavor components (volatile and non-volatile) are protected within a matrix of its components. When the spice is ground, cut or crushed, the cell matrix breaks down and releases the volatile components.

Primary Fuctions of Spices

The flavor, aroma, color or texture that a spice contribute to food or beverage determine its effectiveness in a recipe.Every spice or flavoring contains predominating chemical components that create these sensual properties. A chemical compound of the spice contributes mild to strong flavors. The balance of these compounds gives a spice its own characteristic flavor profile. Spices have characteristic tastes and aromas. They offer six basic taste perceptions: sweet, salty, spicy, bitter, sour and hot. The other descriptive terms include pungent, umami (brothy, MSG or soy source like), cooling, floral, earthy, woody or green. The taste sensations are generally experienced at different locations of the tongue. Sweet is detected at the tip of the tongue, salty at the frontal sides of the tongue, sour at the posterior sides of the tongue, bitterness at the back of the tongue, and heat, depending on the type, at different areas of the tongue.

Table 3: Functions of Spices and Flavorings

1. Taste	Basils, black peppers, cardamom, jalapeno, asafoetida, lemon-grass, star anise, vanilla, kokum.
2. Aroma	Clove, ginger, kari leaf, mint, nutmeg, rosemary, cardamom, tarragon, cinnamon, mango, shrimp paste.
3. Texture	Garlic, mustard, dill, onion, sassafras, sesame seed, shallot, candlenut, almond.
4. Color	Annatto, cayenne, chocolate, paprika, parsley, turmeric, saffron, rose essence, marigold, basil.
5. Antimicrobial	Cinnamon, clove, cumin, oregano, rosemary, sage, thyme, ginger, fenugreek, chile pepper.
6. Antioxidant	Turmeric, rosemary, sage, clove, oregano, mace.
7. Health	Chile pepper, cinnamon, fenugreek, ginger, turmeric, garlic, caraway, licorice, clove.

Table 4: Sensory Characteristics of Spices

Sl.No.	Sensory Characteristic	Spices and Other Flavorings
1.	Sweet	Cardamom, anise, star anise, fennel, allspice, cinnamon.
2.	Sour	Sumac, caper, tamarind, sorrel, kokum, pomegranate
3.	Bitter	Fenugreek, mace, clove, thyme, bay leaf, oregano, celery, ajowan.
4.	Spicy	Clove, cumin, coriander, canela, ginger, thai basil.
5.	Hot	Chile peppers, mustard, fagara, black pepper, wasabi.
6.	Pungent	Mustard, horseradish, wasabi, ginger, epazote, garlic, onion.
7.	Fruity	Fennel, coriander root, savory, tamarind, star anise.
8.	Floral	Lemongrass, sweet basil, savory, turmeric, ginger.
9.	Aromatic	Chervil, black pepper.
10.	Woody	Cassia, cardamom, juniper, clove.
11.	Piney	Kari leaf, rosemary, thyme, bayleaf.
12.	Cooling	Peppermint, basil, anise, fennel.
13.	Earthy	Saffron, turmeric, black cumin.
14.	Herbaceous	Parsley, rosemary, tarragon, sage, epazote, dillweed.
15.	Sulfury	Onion, garlic, chives, asafoetida.
16.	Nutty	Sesame, poppy and mustard seeds.

Some spices have, however, more than a single flavor profile. For example, fennel has not only sweet notes but also bitter and fruity notes; tamarind has fruity and sour notes, while cardamom has sweet and woody notes.

Coloring Properties

Certain spices provide color as well as flavor to foods and beverages *e.g.* saffron, paprika, turmeric, parsley and annatto.

Table 5: Coloring Components Spices

Sl.No.	Spices	Coloring Component	Color
1.	Saffron	Crocin	Yellowish orange
		Crocetin	Dark red
		Beta-Carotene	Reddish orange
2.	Paproka	Carotenoids:	
		Capsanthin	Dark red
		Violaxanthin	Orange
		Cryptoxanthin	Red
		Capsorbin	Purplish red
		Beta-Carotene	Reddish orange
		Lutein	Dark red
		Zeaxanthin	Yellow
3.	Chile pepper	Beta-Carotene	Reddish orange
		Cryptoxanthin	Red
		Capsanthin	Dark Red
		Capsorbin	Purplish red
4.	Turmeric	Curcumin	Orange yellow
5.	Parsley	Chlorophyll	Green
		Lutein	Dark red
		Neoxanthin	Orange yellow
		Violaxanthin	Orange
6.	Annatto	Bixin	Golden yellow
		Norbixin	Orange yellow
7.	Safflower	Carthamin	Orange- red
		Safflor yellow	Yellow

The components responsible for the coloring in spices are oil soluble or water soluble. Some coloring components responsible in spices are *e.g.* crocin in saffron, carotenoids in paprika, capsanthin in chile pepper, bixin in annatto or curcumin in turmeric. The overall coloring characteristic of a spice is sometimes a combined effect of two or more of its coloring components.

Secondary Functions of Spices

Secondary effects of spices are becoming more important in the modern period. Spices can also aid nutrition when they are used instead of salt, fat or sugar to enhance taste in processed foods. Spices may be used in food products as preservaties, which allow for a more natural or friendly label on processed food.

Spices as Preservatives

Spices have long been known for their preservative qualities as (i) antimicrobial and (ii) antioxidants. They have been used by many ancient cultures, Egyptians, Romahs, Indians, Greeks, Chinese and Native Ameriacans, to fumigate cities, embalm the royalty, preserve food and prevent diseases and infections.

Spices as Antimicrobials

Spices were used to preserve meat, fish, bread and vegetables. In Europe, the Middle East and Asia, before the day of refrigeration, they were used alone or in combination with smoking, salting and pickling to inhibit food spoilage. The Romans preserved fish sauce with dill, mint and savory. They preserved meats and sausages with cumin and coriander. The Greeks used garlic to prevent food spoilage. In India, ginger, garlic, clove and turmeric were used to preserve meats and fish. In ancient Egypt, cinnamon, cumin and thyme were used in mummification.

Spices such as cinnamon, garlic and oregano were used to treat cholera and other infectious diseases during Middle Ages. In the late 19th century, clove, mustard and cinnamon were shown to have antimicrobial activity.

Spices have strong, moderate or slight inhibitory activity against specific bacteria. Aldehydes, sulfur, terpenes and their derivatives, phenols and alcohols exhibit strong antimicrobial activity. A combination of spices can be more effective as preservatives than one spice. Gram-positive bacteria are more sensitive to spice than gram-negative bacteria. *Bacillus subtilis* and *B. aureus* are more susceptible than *Escherichia coli* bacteria. Certain spices act as broad-spectrum antimicrobials, such as rosemary and sage, while others are very specific in their functions, such as allspice and coriander. The essential oils of some spices have an inhibitory effect on bacteria and fungi in meats, sausages, pickles, breads and juices. Eugenol in clove sage and oregano; and allicin in garlic are some of the components used as antimicrobials. The more pungent spice and non-volatile oils have also been shown to have strong antimicrobial properties, *e.g.* gingerol in ginger, piperine in black pepper, capsaicin in red peppers and diallyl sulfide in garlic.

Table 6: Antimicrobial Properties of Spices

Sl.No.	Spice	Effective Component	Microorganism
1.	Mustard	Allyl isothiocynate	*Escherichia coil, Pseudomonas, Staphylococcus aureus.*
2.	Garlic	Allicin	*Salmonella typhii, Shigella dysenteriae, molds, yeasts.*
3.	Chile pepper	Capsaicin	Molds, bacteria
4.	Clove	Eugenol	*E. coli* 0157:H7, *S. aureus, Aspergillus, yeasts, Acinetobacter.*
5.	Thyme	Thymol, isoborneol, carvacr	*V. parahemolyticus, S. aureus, Aspergillus.*
6.	Ginger	Gingerone, gingerol	*E. coli, B. subtilis*
7.	Sage	Borneol	*S. aureus, B. cereus*
8.	Rosemary	Thymol, borneol	*S. aureus, B. cereus*

Spices as Antioxidants

Free radicals damage the human cells and limit their ability to fight off cancer, ageing and memory loss. Many spices have components that act as antioxidants and that protect cells from free radicals. Some spices have more antioxidant properties than other depending on the food they are in. Combining spices with other spices or antioxidants such as tocopherols and ascorbic acid produces synergistic effects. The naturally occurring phenolic compounds (phenolic diterpenes, diphenolic diterpenes) in spices are effective against oxidative rancidity of fats and color deterioration of the carotenoid pigments.

Table 7: Spices as Antioxidants

Sl.No.	Spice	Chemical Components
1.	Rosemary	Carnosol, carnosoic acid, rosmanol
2.	Sage	Rosmanol, epirosmanol.
3.	Turmeric	Curcumin, 4-hydroxycinnmoyl methane
4.	Clove	Eugenol
5.	Oregano	Phenolic glucoside, caffeic acid, rosmarinic acid, protocatechuic acid.
6.	Mace and nutmeg	Myristphenone
7.	Sesame oil	Sesaminol, δ-tocopherol sesamol
8.	Ginger	Shogoal, gingerol.

Spices prohibit rancidity and extend shelflife by slowing the oxidation of fats and enzymes. Fats are broken down into peroxides (free radicals) with exposure to air or oxygen and finally into aldehydes and alcohols that give a rancid taste. Spices can halt the oxidative process by blocking or scavenging the free radicals.

Table 8: Therapeutic Effects of Spices

Sl.No.	Spice	Components	Medicinal Property
1.	Turmeric:	(i) Curcumin	(i) Anti-inflamatory, antitumor
		(ii) Curcumene	(ii) Antitumor
2.	Ginger:	Gingerol, Shogoal, Ginge-berane, *Bis*-abilenenan, Curcumine, Gingerol, Shogoal	Digestive aid, stomachache, vomiting, indigestion, stomach ulcers, antitumor, enhance gastrointestinal mobility, inhibit cholesterol synthesis.
3.	Fenugreek:		
	(a) Seed	(i) Trigenelline	(i) Arrests cell growth and prevents hypoglycaemic effect.
		(ii) Diosgenin	(ii) Synthesis of steroid drug and sex hormones.
	(b) Seed & leaf	Soluble dietary fibre (galactomannan) saponins, diosgenin, proptein	Improved glucose tolerance (reduce plasma glucose levels), lowers cholesterol and triglyceride (decreases bile secretion).
4.	Garlic	(i) Allicin	(i) Breaks down blood clots, prevents heart attacks, prevents gastric cancer.
		(ii) Glutamyl peptides	(ii) Lower blood pressure and blood cholesterol
		(iii) Allicin, diallyl sulfide, s-allyl Cystene	(iii) Inhibits platelet aggregation
5.	Licorice	Glycyrrhizin	Treats gastric and duodenal ulcers, prevents coughs and colds, treats chronic fatigue syndrome.

Chapter 5
Medicinal Properties of Spices

Majority of the spices and condiments have curative properties and a profound effect on human health as they affect many functional processes. The natural substances present in them posses the property of curing common human ailments. They serve as powerful natural drugs like antibiotics, carminatives, antidepressants, analgesics, tranquilizers, cholesterol reducers, sex stimulants, antihypertensive, diuretics, anti-inflammatory agents, blood vessel dilators, etc. Some important medicinal properties of spices and condiments are dilated in the following:

Antibacterial Activity

It is noted that bacteria die when exposed to garlic. Garlic was the first antibiotic food that was used against germs. Other spices and condiments with antibacterial activity are clove, cumin seeds, ginger, onion and turmeric.

Anticoagulant Activity

Some spices and condiments such as chilli pepper, garlic, clove, onion and ginger serve as anticoagulants. They discourage platelets, the smallest blood components, from clumping together or aggregating. They are not so sticky and hence build less clots that can clog the arteries. The use of these spices can thus help ward off cardiac problems.

Antidepressant Activity

Certain spices and condiments serve as antidepressants. They help elevate moods by changing brain chemistry *e.g.* cardamom, chilli pepper and garlic. They seem to manipulate mood by influencing serotonin, one of the brains most remarkable neurotransmitters.

Anti-diabetic Activity

Spices and condiments such as cinnamon, curry leaves, fenugreek seeds, garlic and onion possess anti-diabetic activity. Their use can help lower blood sugar or stimulate insulin production in treating diabetes.

Anti-diarrhoeal Activity

Spices and condiments also possess anti-diarrhoeal activity *e.g.* dill, fenugreek seeds, garlic, ginger, mint, nutmeg and turmeric. They help counteract compounds. These spices fight bacteria in the intestines. They thereby exert a soothing effect. They help drain water out the gut and solidify faeces.

Anti-gas Activity

Spices are useful as carminatives-agents that help expel gas and relieve flatulence. The main pharmacological agent is thought to be oils in the plants. These oils relax smooth muscles, thereby allowing gas to escape. In some cases, the gas erupts upward through a relaxed sphincter muscle between the oesophagus and the stomach. Then it is called a belch. Carminative spices also have an antispasmodic, muscle relaxing effect in the intestine. Spices and condiments with carminative activity include aniseed, asafoetida, bishop's weed, caraway seeds, cinnamon, clove, dill, fennel, seeds, garlic, ginger, mint, etc.

Anti-inflammatory Activity

Spices and condiments can reduce inflammation, which is a key process in arthritis and other rhemumatic affliction *e.g.* garlic, ginger, onion, tamarind and turmeric. They help manipulate the prostate gland system to block the complex biochemical inflammatory processes.

Sex Stimulating Property

Some spices and condiments are aphrodisiacal in nature. They help correct sexual inadequancy and dysfunction. They also help build up the health of various sex organs of reproductive system *e.g.* asafoetida, bishop's weed, cardamom, fenugreek seeds, garlic, ginger, nutmeg, onion and pepper.

Calming and Sedative Property

Aniseed, cumin seeds, dill, nutmeg and poppy seeds serve as sedatives and tranquillizers. They work as sedative by stimulating the activity and levels of neurotransmitters such as serotonin that calms the brain.

Mucus-Clearing Activity

Hot, spicy, pungent foods help clear the lungs and breathing passages. They do so by thinning mucus and encouraging it to move along. When a person eats hot food, his eye starts watering and his nose being to run. The same thing happens in the lungs. Hot food activate nerve endings in the oesophagus and stomach, causing watery reactions. Spices and condiments with mucus clearing property are aniseed, asafoetida, basil, bishop's weed, chilli pepper, clove, fennel, garlic, ginger, mustard seeds, onion, tamarind and turmeric. They thin out and help move the lung's secretions. So secretions are coughed out or expelled in a normal way.

Painkilling Activity

People put hot pepper extract on their gums to alleviate toothache. It is the capsaicin in peppers that serve as a local anesthetic and a painkiller. Other spices and condiments which serve as painkilling are asafoetida, bishop's weed, clove, ginger, garlic, mustard seeds, nutmeg, onion and poppy seeds.

Anti-viral Activity

Certain spices and condiments fight various types of viruses, which enter the body through air, water, food, scratches and wounds of skin, and thereby help prevent viral diseases, *e.g.* basil, cinnamon, dill, garlic, onion and turmeric.

Part II

Spices and Condiments

1. Ajowan (Bishop's Weed)
Trachyspermum ammi (Linn.) Sprague

(Syn. *Carum copticum* Heim.)
Apiacae (Umbelliferae)

Ajowan (Ajowain) is an annual herbaceous plant. It bears greysh-brown fruits which constitute the spice. Fruits are 1-seeded, small, ovoid, muricate, aromatic cremocarps, 2-3 mm long, with greysh brown compressed mericarps with distinct ridges and tubercular surface. It is mostly grown in the temperate regions of the world, excepting for a few countries which are cultivated in the tropics, South West Asia (Iraq, Iran, Afghanistan and Pakistan) and especially India and North Africa. In India, it is largely grown in Uttar Pradesh, Bihar, Madhya Prades, Andhra Pradesh, Punjab, Rajasthan, West Bengal, Tamil Nadu and Andhra Pradesh.

Useful Parts
Fruits.

Chemical Constituents
Fruits: Protein, fats, fibre, carbohydrates, minerals: calcium, phosphorus, phytin phosphorus, iron, sodium, potassium; thiamine, riboflavin, nicotinic acid, carotene, iodine, essential oil. Other constituents in the fruits include sugars, tannins and glycosides, besides saponin and steroidal substances.

Processed Products

The processed products manufactured are: (*i*) Oil of ajowan, prepared by steam distillation. (*ii*) Thymol, produced by treatment of the ajowan oil with aqueous alkaline solution and regenerating thymol from it by extraction with either or steam distillation. Both are used in medicine/pharmaceutical industries. (*iii*) Dethymolized oil or thymene for industrial purposes (*iv*) Fatty oil.

Ajowan contains two kinds of oils *viz.*, (*i*) essential oil or volatile oil and (*ii*) non-volatile fatty oil. Two integrated methods are developed to recover both these oils from the crushed seeds. The seed spices are rich in fatty oil.

In Medicine

Seeds are useful in flatulence, colic, atonic dyspepsia, diarrhoea, cholera, hysteria and spasmodic affections of bowels. Seeds produce feeling of warmth and relieve sinking and fainting feelings which accompany bowel disorders. Externally, *ajowan* is applied in mixtures to get rid of rheumatic and neuralgic pains. A tea spoonful of seeds with a little salt is a domestic remedy against indigestion from irregular diet. For stomachache, cough and inigestion, the seeds are masticated, swallowed, and this is followed by a glass of hot water. They are also beneficial in skin diseses. A hot poultic of sseds is used as a dry fomentation to the chest in asthma and to hands and feet in cholera and fainting. They check chronic discharges such as profuse expectoration from bronchitis. Extracts of seeds in 70 per cent and 40 per cent alcohol are toxic to *Staphylococci* and *Escherichia coli*. The leaves are used as a vermicide. Even the roots of ajowan plant are reported to be diuretic and carminative. The aqueous solution of thymol is an excellent mouth-wash and thymol is a constituent of many toothpastes. The distillation water, the essential oil, and the thymol separated therefrom are used in India as medicine, particulary for the cases of cholera.

2. Allspice (Pimenta, Pimento)
Pimenta officinalis Lindl.

Syn: (1) *Pimenta dioica* **(Linn.) Merr,**
(2) *Eugenia pimenta* **DC.**
Myrtaceae

Allspice (Pimenta or Pimento) comprises dried unripe berries from a bushy evergeen small tree, The Name 'Allspice' is derived from the fact that the spice is said to possess the characteristic blend or composite flavour and aroma of cloves, nutmeg, cinnamon and black pepper, all combined in one. The berries are nearly globular, 4-7 mm in diameter, when dried then look-like over-sized pepper berries with a somewhat rough surface and reddish-brown colour. Pimenta is quite different from Pimiento' (*Capsicum annuum*), which is a variety of red pepper. The term 'Pimento' had its origin in the initial incorrect belief of the early Spanish explorers who thought these allspice berries to be similar to pepper. Then came the English

expression 'Jamaica Pepper' and the German name 'Nelkenpfeffer'. It is now officially recognized as 'Pimenta'.

Allspice trees are medium-sized, grow up to a height of 8 to 10 m, and are evergreen. The allspice tree is indigenous to West Indies and tropical Central America, Grenada, Guatemala, Mexico, Honduras and grows semi-wild in Jamaica (W.I.). It is reported to be cultivated in India, especially in West Bengal, Bihar and Orissa. It is also said to grow well and also fruit heavily in Bangalore, Karnataka and Wynad area of Kerala. A few allspice trees are available in Nagarcoil, Kallar and Burliar.

Major Types of Allspice

1. *Jamaican Allspice*: The island of Jamaica is not only the world leader in allspice product, but its allspice traditionally commands premium price. Where the essential oil content of allspice from all sources ranges from 2 to 4.5 per cent, the Jamaican prodct averages 4 per cent. There is only one grade of Jamaican allspice. It is a product of hand-picking and careful sun-drying.

2. *Mexican Allspice*: The Mexican berries are the largest and they are much darker in colour (closer to black than chocolate-brown shade). The Mexican trees are believed

to be of a different strain than Jamaicans. The flavour character of the berries is also different, being less sweet and melowed; essential oil content average is lower than either Jamaican or Guatemalan berries.

3. *Honduran Allspice*: The berries from this origin are similar in appearance to Guatemalans and Jamaicans, although they are larger than the latter. Essential oil content is roughly the same as in Mexicans.

Useful Parts

Berries

Chemical Constituents

Essential oil, quercitannic acid responsible for astringency, a soft resin with a burning taste, fixed oil, proteins, crude starch and traces of an alkaloid.

Whole spice does not lend itself easily to adulteration. The ground spice is some times found adulterated with clove stems and farinaceous products, etc. Pimenta-berry oil is sometimes adulterated with pimenta-leaf oil.

Processed Products

The following products are manufactured from Pimenta:

Pimenta Berry Oil

The essential oil (prmenta berry oil) is obtained by steam distillation of the crushed dried berries for about 10 hours. It has a yellow to yellowish-red colour, darkening with age, and possesses the characteristic colour and flavour of allspice. It contains eugenol as the principal constituent.

Pimenta-leaf Oil

Dried pimenta leaves, on steam distillation yield 0.7 to 2.9 per cent oil which contains eugenol as its main component. It has an inferior odour and flavour to that of the Pimenta berry oil. It is used as a chief substitute for the expensive berry oil.

Pimenta Oleoresin and Oil

Berries are processed on a small scale but little information is available on the precise method of manufacture of oleoresin and oil.

Properties and Uses

The berries are used in a number of preparations *e.g.* as condiment, as a flavouring ingredient in ketchups, soups, sauces, pickles, canned meats, sausages, gravies, relishes, fish dishes, pies, puddings, preserves, etc. They are used in the liquor industry, especially as a favourite ingredient for mullet wines and as perfume in soap-making. It is an important ingredient of whole mixed pickling spice, and spice mixtures, *viz.*, curry powders, mince-meat spice, pasty spice, poultry dressing, frankfurter and hamburger, etc. Ground allspice is delicious in fruit cakes, pies, relishes, preserves, and is good too with sweet yellow vegetables tomatoes. Whole berries are also used in meat broths, gravies and pickling liquids, etc.

In Medicine

Allspice is used as aromatic, stimulant, digestive and carminative. Powdered fruit is administrated in flatulence, dyspepsia and diarrhoea. It is an anodyne against rheumatism and neuralgia.

Berry Oil

There is a good market for berry oil. Pimenta-berry oil is used for flavouring condiments and food products like meat, canned foods, sauces, etc., and in perfumery, soap and pharmaceutical preparations. It is used in the manufacture of soaps, men's spice-based cosmetics such as an after-shave lotion and related ailments. The oil is used as a carminative and stimulant. It shows bactericidal, fungicidal and anti-oxidant properties.

3. Amchur
Mangifera indica Linn.

Anacardiaceae

Amchur is the dried or dehydrated product. It is prepared from unripe but mature, peeled mango flesh, in the form of dried peeled slices or powder. Mango is the most important fruit of India and grows in many states of India. Invariably, un-ripe mangoes or acidic desi or country mangoes are utilized for manufacture of amchur. Amchur is produced commercially in Gujarat, Maharashtra and in most of the northern states of India.

The unripe fruits are peeled and the flesh cut into thin slices. The slices are dried under the sun and packed in gunny bags for sale. *Amchur* is also marketed in the form of 'powder' by crushing or

powdering dried raw mango slices. Sometimes slices are seasoned with powdered turmeric and then sun-dried in order.

Useful Parts

Fruits.

Chemical Constituents

Citric acid, reducing sugars, protein, fats, carbohydrates, mineral matter (Ca, P, Fe), carotene (Vitamin A), vitamin B_2 (riboflavin).

Processed Products

The two important products manufactured are (i) sun-dried peeled mango slices and (ii) mango powder (amchur).

'Amchur' is used as 'souring' agent for curries; similar to the use of tamarind pulp extracts in the South Indian curries, 'sambhar' and rasam. It is also used in chutneys, soups and certain specific vegetable curries. The main purpose of its addition is to lower the pH of gravy, whereby the destruction of the spoilage organisms in the vegetable curry is made much easier at boiling point.

It is further reported that unripe mango is useful in opthalmia and eruptions. The rind is astringent and stimulating tonic in debility of stomach.

4. Anardana
Punica granatum L.

Punicaceae

Anardana comprises dried seeds (with flesh or pulp) of pomegranate. It is a shrub or a small tree 5-8 ft high, a native of Iran, Afghanistan and Baluchistan.

It is found growing wild in warm valley and outer hills of Himalayas, between 900 and 1,800 m and is also cultivated at the selected sites almost throughout India. The pomegranate is an ancient and popular fruit. It is symbolic of plenty and prosperity. It grows best in tropics, below an elevation of 1,000 m, with long, hot, dry summers and cool winters, or in those areas which are continuously warm and dry. High temperature should accompany the ripening season. The fruit is a large, globose berry, shining red, tan, brown, yellowish-green or whitish, when ripe, crowned by the calyx, and is generally 5-7.5 cm in diameter. The fruit is filled with angular hard seeds which are covered with pulp, which, when sun dried or dehydrated, constitute the condiment 'anardana'. 'Anardana' is used as an acidulant in Indian curries particularly in most of the northern states of India.

Commercially Important Types/Varieties

Several types/varieties are cultivated. These differ in shape and size of fruit, colour and thickness of rind, colour of aril and nature of seeds, degree of sweetness and/or acidity, etc.

(I)   Varieties Important in India

1. *Alandi*: Fruits are medium in size, with blood-red or deep-pink arils and with sweet but slightly acidic pulp. Seeds are very hard.

2. *Dholka*: Fruits are large; rind greenish-white; arils are fleshy and white or pinkish white and sweet soft.

3. *Ganesh (GBG1)*: It is a selection from seedling progenies made by Dr. G.S. Cheema (Late), Horticulturist, erstwhile Bombay State. It is an excellent, high-yielding selection with fruits which are medium large to large; arils are pink coloured; seeds are soft and high in total soluble solids (TSS). This is the most important commercial variety.

4. *Bessein Seedless*: Fruits are medium sized, with greenish-yellow skin. Arils are pinkish white. Seeds are soft.

5. *Joth (GKVK)*: The fruits are medium to large in size (220 g), have attractive yellowish colour with deep-pink arils. Seeds are soft. It has high TSS and high pulp content (70.5 per cent).

(II)   Varieties Important in Other Countries

1. *Kabul*: Fruits are large; rind is thick and deep-red coloured mixed with pale-yellow; arils are dark (blood) red and juices slightly bitter.

2. *Kandhari*: Fruits are large, rind is deep-red; arils are flesh and hard blood-red, slightly acidic but juice is sweet; seeds are hard.

3. *Selimi*: (from Baghdad) Bears very big fruits weighing 1 kg each, the fruit is juicy and sweet.

4. *Roman Chokab*: Black pomegranate (from Baghdad) bears fruit of dark colour with large seeds. It has a tender sweet-sour taste.

5. *Wellissi*: (from Palestine) Bears large, sweet and early fruits.

6. *Ras-el-Baghi*: (from Palestine) Bears large, sweet and early fruits.

7. *Suffami*: (from Palestine) Produces big juicy fruit with delicious sweet-sour taste. The best specimen of this variety is obtained from Kabul, Afghanistan and Perisa.

8. *Dulca Colorado*: Fruit big, sweet and tasty.

9. *Grand Blanc*: (Spain) Fruit seedless.

10. *Choodesuy*: Fruit is of very high quality; stands transportation well.

11. *Boomageny*: (USA) Fruits are large, very juicy and stand shipment well.

Useful Parts

Fruits.

Chemical Constituents

Fresh Fruits (Seeds with flesh):

Carbohydrates, mineral matter (Ca, Mg, P, Na, Cu, S), oxalic acid, phosphorus, iron, sodium, potassium, copper, chlorine, vitamin B_1 (thiamine), vitamin B_2 (riboflavin), nicotinic acid and vitamin C.

Processed Products

The pomegranate juice is extracted from its fresh flesh seeds. The juice is used for preparing 'Anar Sharbat' or 'syrup'.

Propertis and Uses

Anardana is motly used as a condiment for acidification of chutneys and certain curries. The rind is also used as dye in material for cloth.

In Medicine

The seeds are stomachic while the pulp is both cardiac and stomachic. The decoction of the fruit combined with cloves, etc., is useful in diarrhoea and dysentery.

5. Aniseed
Pimpinella anisum Linn.

(Syn. *Anisum vulgare* Gaertner,
***Anisum officinalis* Moench)**
Apiaceae (Umbellifereae)

Aniseed or anise is a popular spice used midely throughout the world. It is also known by different names, *anysum* by early Arabs, anisen by Greeks and then anise by the English. Ancient Assyrians

used it as a medicine. Greeks used it as digestive aid. Romans used to soothe sore throats. It was also used as an aphrodisiac and as a charm against mightmares. Being associated with the taste of licuorice (Jesthamadh), the orthuguese called it 'erva doce' (sweet herb) and the Indonesians called it 'jintan manis' (sweet seeds). It is an annual herbaceous plant. It is a native of the East Mediterranean region. It is widely cultivated in Bulgaria, Cyprus, France, German, Italy, Mexico, South America, Syria, Turkey and Russia. In India, It is grown to a small extent as a culinary herb or as a garden plant. Anniseed is believed to have been introduced in India by the Mohammedan invaders from Persia.

In European countries, aniseed is sometimes mistaken for another spice 'star-anise' (*Illicium verum*) which is a tree-spice Aniseed bear fruits which contain similar volatile oil. They have hence almost similar odour and flavour. However, anise oil is finer and has a characteristic agreeable odour and a pleasant aromatic taste. Aniseed is available whole or ground. Star-anise is slightly cheaper than the aniseed.

Useful Parts
Seeds.

Chemical Constituents

Protein, fatty oil, essential oil, sugars, starch, N-free extract, chlorine, minerals (K, Fe, P, Ca), anise oil (anethole, methyl chavicol, p-methoxphenyl acetone, terpenes).

Processed Products

Aniseed Oil

Aniseed yields an essential oil on steem distillation. It is known as 'oil of anise'. It is usually replaced by the fruit for medicinal and flavouring purposes. Anise oil is colourless to pale-yellow liquid with the characteristic odour and taste of the fruit. Exposure of the oil to air causes polymerisation, and some oxidation also takes place with the formation of anisaldehyde and anisic acid.

Propertis and Uses

Aniseed is used for flavouring curries, food, sweets, cookies, biscuits, cheeses, liquors like arraes, pernod, sambuca, bakery products, beverages and anisette etc. Essential oil is used in perfumery, soaps and other toilet articles and for flavouring culinary preparations, confectionery, beverages and liquor anisette. It is used in perfuming sachets, dental preparations and mouth washes. It is also used in the manufacture of liqueurs. The residue left after extraction of oil may be used as a high-grade cattle feed.

Ether extraction of the exhausted fruits yields a dark green fatty oil suitable for soap making. A hard fraction of the oil can be used as a substitute for cocoa butter in confectioneries and pharmaceutical preparations. The fatty oil expressed from the whole fruits possesses the characteristic anise flavour.

Fresh leaves of the plant are used as a garnish and for flavouring salads. They are eaten as pot-herb as they contain an essential oil and vitamin C.

In Medicine

The fruit is a mild expectorant, stimulating, carminative, diuretic and diaphoretic. It is used in flatulent colic, in the preparation of asthma powders and in veterinary medicine. Oil of anise is used as an aromatic carminative to relieve flatulence. It is used as an ingredient of cough-lozenges in combination with liquorice. It is a mild expectorant. It is also used as an antiseptic and for treating cholera. Essential oil is also used externally as an insecticide against small insects such as head-lice, mites and vermin.

Aniseed is beneficial in cataract. Infusions of seeds and honey are useful in preventing gas and fermentation in the stomach and bowels. A tea made from this spice is used to calm nerves and induce sleep. Aniseed tea is helpful to promote milk of breast.

6. Asafoetida (Asafetida)
Ferula asafoetida Linn.

Apiaceae (Umbelliferae)

Asafoetida is the dried oleo-gum or gum oleoresin. It is exuded as a very thick and sticky and pasty sap or latex from the living underground rhizome or rootstock of several species of *Ferula*. These species are the perennial herbs, 1-1.5 m high. They are the sources of the oleo-gum resins, used as condiment as well as in medicines. The species are distributed from the Mediterranean region to Central Asia.

Its main growing areas are the eastern parts of Iran and the western part of Afghanistan.

There are about 60 species of the genus *Ferula* which are widely distributed in three main regions, *viz.* (i) Central Asia (ii) Europe and (iii) North Africa. *Ferula* grows from perennial rootstocks and attains annually a height of 1 to 3 m. Some commercially important plant species are: 1. *F. aliacea* Boiss., 2) *F. asafoetida* Linn., 3) *F. communis*

Linn., 4. *F. foetida* (Bunge) Reget 5. *F. ferulago* Linn., 6) *F. galbaniflua* Boiss.*et* Buhse,7) *F. hermoins* Boiss, 8) *F. jaeschkeana* Vatke, 9) *F. marmarica* Asch. *et* Taub, 10) *F. narthex* Boiss., 11) *F. orientalis* Linn., 12) *F. persica* Willd, 13) *F. rubricaulis* Boiss, 14) *F. schair* Brosz., 15) *F.sumbul* Hook.*f.*, 16) *F.szowitziana*, 17) *F.tingitana* Linn. (syn. *F. sancta* Boiss.).

Types of Asafoetida Imported into India

The more important ferula oleo-gum resins imported into India, chiefly from Iran and Afghanistan are: (a) asafoetida, (b) galabanum and (c) sumbal. Of these, asafoetida is the most important one. A part of the imported resin is re-exported from India.

Useful Parts

Oleo-gum resin.

Chemical Constituents

Asafoetida contain, resin, gum, volatile oil, ash, asaresinotennol free or combined with ferulic acid, Umbelliferone.

Processed Products

1. *Oil of Asafoetida*: The oil of asafoetida is obtained by the steam-distillation of gum-resin. The chief constituent of

the oil is secondary butyl propenyl disulphide. The remaining constituents are other disulphides, trisulphide, pinene, another terpene and an unidentified compound. The disagreeable odour of the oil is due to the disulphide ($C_{11}H_{20}S_2$). The flavouring and pharmaceutical industries chiefly use alcoholic tinctures of the gum-resin.

2. *Oil of Galabanum*: Galabanum is the oleo-resin of *E. galbaniflua*. It is obtained from its roots. It is yellowish to brownish. It becomes liquid on heating. It yields an essential oil.

Propertis and Uses

Asafoetida is extensively used in India to flavour curries, sauces and pickles. In Iran (Persia), the natives rub asafoetida on warmed plates prior to placing meat on them. Besides, the large cabbage-like tops of the plants are relished raw by the natives.

In Medicine

It is stimulant, carminative, expectorant and antispasmodic. It is used as uterine tonic. It is useful in asthma, whooping cough and chronic bronchitis. It is used as an enema for intestinal flatulence. It is also advised in hysterical and epileptic affections and in cholera. It is a very effective remedy for relieving spasms and in indigestion and colic. It is beneficial externally on the stomach to stimulate intestines. It is useful as a plaster on inflammatory swellings.

It is a nervine stimulant, digestive agent and sedative. It excites the secretion of ovarian hormones and sex stimulating centers. It expels wind from the stomach. It is beneficial in spasmodic disorders, indigestion and colic. It is useful in fevers, amnesia, hysteria, aches and pains, leucorrhoea, excessive menstruation, etc.

7. Balm (Lemon Balm)
Melissa officinalis Linn.

Lamiaceae (Labiatae)

Balm (Lemon balm) is a perennial herbaceous plant. It is called lemon balm because of strong agreeable odour, reminiscent of lemon. It is evergreen, 30-60 cm in height. The leaves or the herb is used as a spice and as a flavourant. A native of the countries bordering northern Mediterranean, it grows wild and is also cultivated in

gardens as a medicinal herb. It has been naturalized in the USA, grows wild in the eastern USA, and is found in the temperate Himalayas. Another plant *Melissa parviflora* Benth., considered a good substitute for *M. officinalis* (Lemon Balm), is found in temperate Himalayas from Garhwal to Sikkim, Darjeeling and Khasi, Aka and Mishmi hills, at an altitude of 1,200-3,000 m. Leaves are ovate or ovate-lanceolate. Balm is now well established in Himachal Pradesh (Solan).

Processing Technology
Essential Oil

The volatile oil is obtained from fresh herb. By distilling fresh *Melissa officinal* -is at the beginning of the flowering stage, could obtain 0.014 per cent of the true oil. The fresh herb during full-bloom give 0.010 per cent of oil. The odour suggested presence of citral and citronella; the scent of the first mentioned oil was more agreeable, typical of Melissa.

Useful Parts

Entire herb, leaves.

Chemical Constituents

The main chemical components are caffeic acid (a tannin), several flavonoids (luteolin-7-O-glucoside, isoquercitrin, apigenin-7-O-glucoside, and rhamnocitrin), rosmarinic acid, ferulic acid, methyl carnosoate, hydroxycinnamic acid, and 2-(3',4'-dihydroxyphenyl)-1,3-benzodioxole-5-aldehyde. The essential oil (0.19 per cent v/w; obtained by a water distillation method) from the flowers contains various aldehydes: geranial (major constituent in oil), citronellal, beta-caryophyllene, neral, and geranyl acetate.

Steam distilled from tops and tender green leaves produces small quantity of oil up to 0.02 per cent. Mainly Geraniol, linalool, with citronellol and citronellal.

Propertis and Uses

Balm leaves are widely used for culinary flavouring. Fresh or powdered balm leaves are used in fish dishes stufings or as a substitute for lemon. Fresh leaves are piquant in salads and in summer drinks. The volatile oil distilled from plant is used for flavouring and also somewhat in perfumery.

In Medicine

Leaves and flowering tops are medicinally useful. Lemon balm is stomachic, anti-tubercular and has anit-pyretic properties. It strengthens gums and removes bad taste from mouth. The fruit is a brain tonic and is useful in hypochondriac condition. Leaves and stems are said to be useful for liver and heart diseases and also in bites of venomous insects. The Balm oil is reported to be sedative. Balm distillates can act as mild spasmolytic agents.

8. Basil (Sweet Basil)
Ocimum basilicum Linn.

Lamiaceae (Labiatae)

Basil (French Basil, Sweet Basil, Niyazbo) is an erect glabrous herb, 30-90 cm high, native of North-Western India and Persia. It constitutes an important culinary herb. It is a rich source of a valuable essential oil. In India, there is wide spread belief that if planted around homes and temples, it ensures happiness. Hindus consider it sacred and also good for health when its fresh leaves are taken raw and also their decoction. This delightful annual herb of the mint family is native to India and Persia. In France, it is called the

'herbe royale' and its aroma is highly esteemed in that food-loving
nation. It is indigenous to lower hills of the Punjab and Himachal
Pradesh. Apart from India, it is cultivated in Southern France, Egypt,
Belgium, Hungary and other Mediterranean countries and also in
the USA. The plant grows to a height of about 90 cm. The freshly
picked bright green leaves measure up to 3.75 cm in length. When
dried, they turn brownish-green, whole and broken brittle, curled or

folded together. Dried leaves and tender 4-sided stems of this plant are used as a spice for flavouring and for recovery of essential oil.

The leaves have numerous dot-like oil glands in which aromatic volatile oil of the herb is contained. It can be easily grown at home or in gardens in ordinary soil.

There are different types of Basil *viz,* American Basil, French Basil, Egyptian Basil and Indian Basil.

There are numerous varieties of *O. basilicum,* of which four identified in India, are : (1) var. *album* Benth. (lettuce-leaf basil); (2) var. *differme* Benth. (curly-leafed basil; (3) var. *purfurascans* Benth. (violet-red basil) (4) var. *thyrsiflorum* Benth. (common white basil).

Useful Parts
Leaves.

Chemical Constituents
Protein, carbohydrates, volatile oil, fixed oil, cellulose, pigment, vitamins. Vit. B_1(thiamine), niacin, Vit. B_2 (riboflavin), vit. C (ascorbic acid), and vit. A, minerals (Ca, P, Na, Fe), linalool, ocimene, methyl cinnamate, methyl chavicol.

Processed Products
The main product manufactured from basil leaves and flower tops are the essential oil known as 'Oil of Basil'.

Distillation of Oil
Oil of sweet basil is produced by the hydro-distillation of the herb. The flowers or whole herbs or both are packed into a distillation unit and hydro-distilled or steam distilled. The oil, being lighter than water, is easily separated from oil-water mixture. It is advisable to use distillate after removing oil for further charge, as it contains small quantity of oil. Oil yield thus increases. The oil obtained from flowers is better than the oil from whole herb in quality.

Properties and Uses
Sweet basil is used for flavouring in a variety of culinary preparations *e.g.* cheeses, tomato cocktail, eggplant, zucchini, cooked cucumber dishes, cooked peas, squash and string beans. Chopped basil is sprinkled over lamb chops before cooking. Basil is often used with, or as a substitute for oregano in pizza topping, spaghetti sauce or macaronic and cheese casseroles. It is also useful in the

manufacture of Chartreuse and other liqueurs. Basil is an important seasoning in tomato-paste products in Italy.

The oil of sweet basil is also employed quite extensively in all kinds of flavours, including those for confectionery, baked goods and condimenatory products (chilli sauces, catsups, tomato pastes, pickles, fancy vinegars) and in spiced meats, sausages etc. The oil also serves for imparting distinction to flavours in certain dental oral products and in certain perfumery compounds. The lower-priced Reunion oil is preferable for the scenting of soaps.

In Medicine

Sweet Basil is stomachic, anthelmentic, alexipharmic antipyretic, diaphoretic, expectorant, carminative, stimulant and pectoral. An infusion of the plant is advised for cephalalgia and gouty joints, and used as a gargle for foul breath. The juice of the leaves is useful in the treatment of couph and coughs. It is used as a nasal douche, as a nostrum for ear-ache and also for ring-worm. Seeds possess demulcent, stimulant, diuretic, diaphoretic and cooling properties. They are given internally in cases of habitual constipation and piles. Root is used in bowel complaints in case of children. Flowers are carminative, diuretic, stimulant and demulcent as well.

9. Black Pepper
Piper nigrum Linn.

Piperaceae

Black pepper is regarded as the 'King of Spices'. Its centre of origin is the Western Ghats of India. Since time immemorial, India has been the largest supplier of black pepper. Presently, it is also grown in the countries like China, Vietnam, Thailand, Malaya, Indonesia, Sri Lanka, Madagascar, South Africa, Mobambique, Cameroon, Brazil and Mexico. In India, it is being cultivated in the states of Karnataka, Kerala, and Tamil Nadu.

The generic name *Piper* is derived from Indian names for plant *e.g.* Pippul (a Bangalese name), Pipuli or Pippati (Sanskrit name). The specific name *nigrum* owes to its black seeds. Black pepper are the fruits of evergreen woody perennial vine indigenous to Malbar coast of India. It was important commodity of trade between Europe and the Orient. The magic of the pepper and other spices attracted European voyagers like Columbus, Vasco Da Gama, Drake and few others to reach India during the last quarter of 1st century. In India, it is grown as a homestead crop in every compound, a mixed or intercrop trailed on various trees, a pure crop on slopes or a mixed crop on shade trees in tea and coffee plantations. Black pepper is also traditionally regarded as 'black gold', which can be kept, apart for providing liquid cash when most needed.

There are many varieties of black pepper known in the world trade *e.g.* Tellichery, Malabar, Alleppey (Kerala), Lampang, Saiggon, Penang and Singapore. These are names derived from the localities where grown or from ports through which they are being exported. These vary slightly in their physical and chemical characters like colour, size, shape, flavour and bite. Tellichery, Alleppey peppers and Malabar Garbled (MG) are large, attractive, dark reddish-brown to black, very aromatic and are among the best. Lampang and

Singapore peppers are smaller, more shrivelled, however, almost equally pungent. Following are the popular cultivated varieties:

1. *Arakkulamunda*: It is moderately good and regular bearer from the central area of Kerala. It has derived its name from a village Arkkulam in the Thodupucha taluk of Idukki district. It is commonly found in the settlement areas of north Kerala. It yields regularly and matures earlier than most other varieties. The spikes are medium long and berries are bold and heavy. It yields 9.8 per cent oleoresin, 4.4 per cent piperine and 4.7 per cent essential oil.

2. *Balancotta*: It is found in north Kerala and grows very vigorously. It bears the largest leaves among the Kerala varieties. The vines are large. The spikes are medium long to long, setting moderately good, and berries are bold, pale-green. Early to medium duration berries give about 30 per cent driage. It gives 9.3 per cent oleoresin, 5.1 per cent essential oil and 4.2 per cent piperine. Its oil content is one of the highest and has very high flavour quality.

3. *Cheriyakaniyakkadan*: It is a popular type, bears regularly, yield heavy and of high quality, wilt resistant; popular in north and central Travoncore (Kerala). Leaves are small, elliptic; spikes are of medium length, closely set with medium-sized, dark green fruits.

4. *Cheriakodi*: It is popular in north Travancore and north Malabar (Kerala). Leaves are narrow, dark green; spikes are short, with dark or pale green fruits which are the smallest among all types. Plant is dwarf and sturdy type, bearing in alternate years; quality is high. It yields 38 per cent dried pepper.

5. *Daddagya*: It is popular in north Kannara (Karnataka). Leaves are broad and spikes are long, curved. The fruits are large among Mysore types. It is uniform yielder, reputed for making white pepper; yield 38 per cent dried pepper.

6. *Kalluvally*: It is a promising north Kerala cultivar. Its distribution is rather restricted to settlement areas of the submontane pepper tracts of Wynad and Cannanore districts. It is hardy and regular yielder. Leaves are

medium, ovate, elliptic and spikes are medium and have got a characteristic twisting due to very thick setting. Berries are small to medium, heavy and have high driage (about 40 per cent). It is a regular bearer and reportedly tolerant to water stress and diseases.

7. *Kaniakkadan*: The name appears to have been derived from Kaniakkar, a tribe inhabitating the hilly Western Ghat areas of present –day Idukki district. There are four different 'Kaniakkadan' types. These are 'Cheriyakaniyakkadan', 'Valiakaniakkadan', 'Karutha Kaniakkadan' and 'Valutha Kaniakkadan'. 'Cheriyakaniyakkadan' is a popular cultivar of eastern parts of Kottayam and Quilon districts. It is also found commonly in many areas of north Kerala. It is bisexual type, having small elliptical leaves; medium, long spikes and medium-sized berries. It is an average yielder and a regular bearer. 'Valiakaniakkadan' has got slightly larger leaves, longer spikes, and bold and heavy berries. Its berry setting is good. It is moderately high yielder.

8. *Karimcotta*: It is popular in north Travancore and north Malabar (Kerala). Leaves are large, dark green and spikes are short, curved with closely set large, dark green fruits. It is hardy, regular bearer and good yielder.

9. *Karimunda*: The name might be due to bluish-black tender shoots and dark-green leaves and berries. It is the most popular cultivar grown throughout Kerala. It is bisexual and have small more or less oval leaves and short to medium long, closely set spikes. The spikes are 4 -10 cm long or even more in certain cases, with a mean leanth of around 6.5 cm. In central Kerala, most of vines have short, well-filled spikes, but the spikes found in the Idukki district and also in parts of Wynad are much longer. It is prolific and regular bearer, having medium sized berries of good driage and yield good quality pepper.

10. *Kottanadan*: It is the most popular pepper in the south Kerala. This is a vigorous growing, bisexual type having large, broad, ovate leaves, long spikes, high fruit set and medium-sized berries. The cultivar is high and regular yielder. It gives about 37 per cent driage and produces

high quality, heavy pepper. In Wynad areas, 'Kottanadan' is grown as 'Aimpiriyan'; the name is derived from the fact that berries are arranged in five distinct rows on the spike. 'Kumbhakodi', a cultivar grown in certain areas of Quilon seems to be a variant of 'Kottannadan' or 'Aimpiriyan'.

11. *Kuthiravally*: The name is derived from its spikes being horse-tail like. It is found in many pepper growing areas throughout Kerala. It is a moderately high yielder, but alternate bearer. The ovate leaves are medium large, spikes very long and slender and flowers are bisexual. Berries are medium large, having high driage. 'Kuthiravally produces high quality pepper.

12. *Malligesara*: It is common cultivar of Uttara Kannada of Karnataka. It is moderate yielder, having medium large leaves and spikes. Two types of 'Malligesara' are known, 'Karimalligesara' and 'Bilimalligesara'. These are differentiated based on anthocyanian colouration of emerging shoots. In 'Karimalligesara' it is purple-white and in 'Bilimalligesara' it is pale-green.

13. *Narayakkodi*: It originally hails from Kottayam-Champakkara-Mallappally tract, but is presently found throughout Kerala. It is a regular average yielder, having high driage. The leaves are small to medium, ovate and have got a characteristic twist in leaf blade. Plant is bisexual, having small spikes with a characteristic twisting that results from thick berry setting. The persistent stigmatic base gives the feeling of a 'pin' and hence the name 'Narayakkodi' (Narayam- a long iron – nail like tool used in olden days for writing on palm leaves).

14. *Neelamundi*: It is a popular especially in the Kattappana and Kallar areas. It is a moderately good yielder, having medium- large leaves and vigorous growth; medium long spikes and thickly set medium sized berries.

15. *Panniyur* I: This is a hybrid pepper variety developed from across between 'Uthirankotta' (female) and 'Cheriyakaniyakkadan' (male) at the Pepper Research Station (KAU), Taliparamba. The plant is vigorous and has cordate leaves. Spikes are long, berries bold and filling

is good. It is a prolific bearer in open areas. It is also an early bearer, having medium maturity. This variety is, however, not very suitable for intercropping. In shade and also in heavy fertility conditions, the flowering and yield seem to come down and there is more of vegetative growth.

16. *Uddaghere*: It is a good yielder, regular bearer and a very popular cultivar of Uttara Kannada. The leaves are medium large and spikes long with good settings. It has high driage (39-40 per cent) and produces good quality pepper. The cultivar is commonly intercropped and is found to give yield.

17. *Valuthanamban*: As the very name indicates, the young emerging shoot and leaves of this cultivar have a whitish tinge, which at maturity turns dark green. It is another moderate yielder found in pepper tracts of Idukki. It has medium-long spikes. Berries are closely set on spike and moderately heavy.

Useful Parts
Berries (fruits).

Chemical Constituents
Essential oil, oleoresin, piperine (alkaloid), apart from chavicine, piperidine and piperettine.

Properties and Uses
Its value as an essential preservative for meats and other perishable foods is known over centuries. It is, therefore, largely employed by meat packers and in canning, pickling, baking confectionery and preparation of beverages. A final dash of pepper is used effectively to adjust the flavour just before the end of cookings.

Black pepper constitutes an important component of culinary seasoning of universal use and an essential ingredient of numerous commercial food-stuffs. It is an important constituent of whole pickling spice and many ground spice formulae of seasonings etc. for poultry dressings, sausages, hamburger and frankfurter seasonings.

White pepper is used in mayonnaise. Oil is a valuable adjunct in flavouring of sausages, canned meats, soups, table sauces and certain beverages and liquors. It is used in perfumery, particularly in bouquets of the oriental type to which it imparts spicy notes difficult to identify. The oil is also used in carnation compound for soaps. It possesses some feeble anti-periodic property, however, is no longer used in medicine. Piperine has been used to impart a pungent taste to brandy. It has also been tried as an insecticide. Oleoresin is used for flavouring of sauces, sausages, chutneys etc. as it renders the processed foods of uniform quality and flavour.

Processed Products

Followings are value added products of black pepper:

1. *White Pepper*: White pepper is flavoured because of its mellow flavour, mild pungency, low fibre and high starch content and above all, white colour and absence of black particles. In the importing or user countries, the USA, Germany, UK, the white pepper is ground into powder and packed in retail consumer packings. In countries like Indonesia, white pepper is made either by removal of skin of ripened berries or by retting and abrasion peeling of black pepper berries.

2. *Canned Tender Green Pepper*: Canned green pepper finds use in Federal Republic of Germany, France, Belgium, Finland, Denmark, the USA, Japan and Middle-east countries as a garnishing spice and is preferred to black pepper due to freshness, natural attractive green colour, aroma and mild pungency.

3. *Bottled Green Pepper in Brine/Vinegar/Acetic Acid*: Berries and spikes packed in 2 per cent brine containing 2-4 per cent aceitic acid are well preserved. Air-tight capping of bottles prevents mould attack during prolonged storage. Berries and whole spikes could be preserved in 16 per cent brine containing only 1-2 per cent acetic acid. Berries or spikes could also be preserved without any diifficulty in 2-4 per cent acetic acid solution alone without the use of common salt but is affected natural green colour of pepper. It turns brown but taste and flavour remains good.

4. *Bulk Packed Green Pepper in Brine*: Bulks packing of whole spikes in PVC containers containing 16 per cent brine and 1-1.5 per cent of acetic acid solution have given encouraging results. Alternatively, use of 16 per cent brine

containing 0.25 per cent citric acid and 100 ppm SO_2 helps in better retention of colour of green pepper.

5. *Other Bottled Green Pepper Products*: Different types of mixed pickles using green pepper as one of the components prepared in brine, vinegar and oil remain in good condition for about six months. A green pepper is advantageously incorporated in preparation of rasam, soups etc. Canned green pepper rasam remains in good condition for six months.

6. *Cured Green Pepper (Dry-packed)*: Cured green pepper packed in moist conditions in flexible packages and without any covering liquid has been developed to overcome disadvantages of poor texture and weaker flavour of dehydrated green pepper and heavy weight and high packaging cost of canned and bottled green pepper.

7. *Dehydrated Green Pepper*: The pungent fragrance of the dehydrated green pepper is universally liked in meat processing and sausage-manufacturing. Dehydrated pepper is now being preferred because of its light weight and hence much lower packing cost and freight than canned or bottled green pepper, though its flavour is rather less and texture harder.

8. *Freeze-dried Green Pepper*: It is far superior to both canned or bottled dehydrated types of green peppers because of its superior attractive green colour, better aroma and texture. However it is more expensive than dehydrated green pepper because of costly sophisticated processes used. Such pepper on thawing will be almost equal to fresh material.

9. *Frozen Green Pepper*: The cleaned, blanched, cooled berries are frozen in 2 per cent brine containing 0.25 per cent citric acid so as to give a drained weight of 55 per cent. Ascorbic acid (vitamin C) is added to the covering brine before freezing, canning or bottling, to serve as an antioxidant for prevention of browning in green pepper and brine.

10. *Pepper Powder*: Fully mature black pepper graded as MG1 is used for preparing powder.

11. *Oil of Pepper*: This is prepared by distillation of powdered pepper. The quality and pepper composition of pepper oil varies considerably; generally due to variation in variety, grade, storage conditions and processing. To arrive at uniform quality in pepper oil, a proper blending of inputs is necessary. The characteristic aroma of pepper is due to the volatile oil present. Generally, slightly immature pepper will have more oil.

12. *Pepper Oleoresin*: It is prepared by solvent extraction of ground pepper. It is the concentrate of all the flavour components (aroma, taste, pungency, etc.). It is extracted from black pepper using solvent like acetone, ethylene dichloride, etc.

13. *Pepper By-products*: Three different by-products are available in the market *viz*. (i) the pepper rejections or waste, (ii) 'varagu' or unfertilized buds, and (iii) stems and inflorescene stalks. Varagu and stalks are poor in ether soluble fractions and have a high content of crude fibre. Pepper rejections, however, are rich in the bite factor and can be used for the preparation of oleoresin in order to economize the use of pepper as a condiment and replace it in times of scarcity.

14. *Pepper Hulls*: Pepper hulls or shells removed during the preparation of white pepper are sold separately as a light to dark-brown powder with a very pungent odour and taste, and found useful for flavouring tinned foods. Pepper shells (freshly removed) are rich in volatile oil and can be used as a source of pepper oil.

In Medicine

Black pepper is a stimulant, pungent, aromatic, digestive and nervine tonic. The ancient Aryans considered pepper as a powerful remedy for various disorders such as dyspepsia, malaria, delirium tremors, haemorrhoides, etc. The Asians are said to have used it as an aphrodisiac. It is an appetizer and good home remedy for digestive disorders. It is beneficial in cold and fever. It benefits in various ailments such as amnesia, cough, prolapse of the rectum, snake-bite, impotence, tooth-ache, muscle pain, etc.

10. Caper
Capparis spinosa Linn.

Capparidaceae

The word 'caper' is originated from the Latin word capra (means goat). The name reflects its strong smell. It is considered to originate from the near East or Middle East. Capers are the buds of the unopened flowers of *Capparis spinosa* Linn. It is a low bushy spiny shrub with close foliage. It grows in the south of Europe, North Africa and in India in the low inner valleys of Himalaya, Chamba, Kumaon, Maharashtra, Konkan, Andhra Pradesh (Deccan Peninsula), Western Ghats, Punjab, Rajasthan and

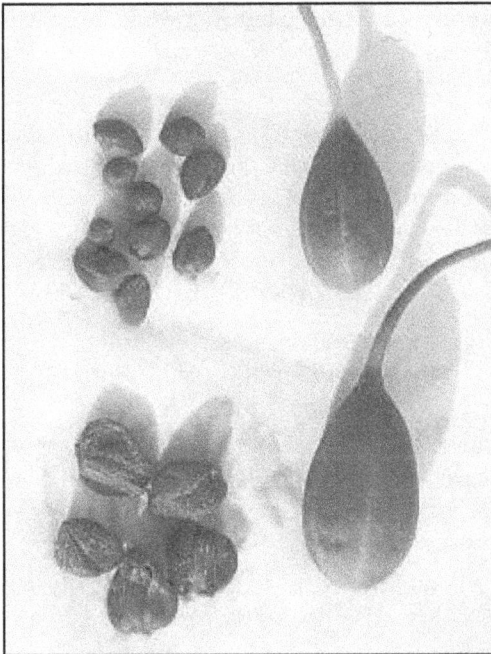

north-western India. It is valuable mainly for its flower buds which are picked and sold as 'capers'. It is deciduous with roundish leaves. The white flowers have purple-tipped stamens.

The tiny green buds open when the sun rises and close when it sets. Once cut, they remain closed. They are cured and prepared in salt.

Useful Parts
Flower-buds

Chemical Constituents
Flower-buds: glcoside rutin which on acid hydrolsis, gives rhamnose, dextrose and quercetin and glucocapparin, rutic acid, pectic acid, a substance with garlic odour, a volatile emetic constituent and saponin.

Propertis and Uses
Capers are unequalled in flavour and used most expertly by many European and American cooks. Fish and meat sauces rendered delicious with addition of few capers. It is used as a garnish for cold roasts and salads. Capers are also used for flavouring pickles and relishes. Caper sauce with boiled mutton is a great favourite in Western countries.

In Medicine
In India, caper buds (as well as the fruits) are thought useful in scurvy. Capers have been used to aid digestion to prevent diarrhea and to increase appetite.

11. Caraway and Black Caraway
Carum carvi Linn.

Carum bulbocastanum W. Koch,
Syn. *Apium carvi* (Linn.) Crantz.
Apiaceae (Umbelliferae)

Both caraway (*Carum carvi* Linn) and black caraway (*Carum bulbocastanum* W. Koch) have been described in Hindi as '*Siha zira*','*Diya zira*' or '*Kala zira*'. Both species are and covered under one similar head.

Caraway is mentioned in the 'Ebers papyrus' (1552 BC) an Italian medicinal treatise. Its seeds were found in a pile of 5000

years old debris left by primitive Mesolithic lake dwellers in Switzerland. It is mentioned in the 12th century German medical book and a 14th century English cookbook. It was useful extensively by ancient Greeks and Romans and also widely used in the middle ages. About 2000 years ago, Roman soldiers dispersed caraway seeds from its home in Asia Minor to most of the known world. The English and Germans have been using it for many centuries. It is being used since long in India as well. Caraway is the fruit of a biennial herb. It is a native of North and Central Europe. It is extensively cultivated in Holand, Russia, Poland, Bulgaria, Denmark, Rumania, Syria, Morocco and to a small extent in England. It has been introduced to the USA, where it is grown as a garden-crop. It (black caraway or *Kala zira*) also grows wild in the north Himalayan regions. It is also cultivated in the plains as a cold season crop. However, in the hills of Kashmir and Kumaon it is cultivated at an altitude of 2,740 to 3,660 m as a summer crop. The caraway plant attains a height of 0.5to 0.6 m. The small white flowers are borne in compound umbels. The fruit, when ripe, splits into narrow, elongated carpels. It is 4 to 6.5 mm long, curved, pointed at the ends, and with 5 longitudinal ridges. The dried fruit is brown in colour, has pleasant odour and aromatic flavour. Seeds are hard and sharp to touch. Black caraway has been found growing wild in the districts of Kinnaur, Lahaul and Spiti and tehsil Pangi and Bharmour of Chamba district. It is important as a spice and more as a valuable seed in Ayurvedic medicines. It, being a perennial crop, does not require planting year after year. Low rainfall during the vegetative, flowering and maturity stages, helps in developing flavour and quality of the seeds.

Useful Parts

Seeds (fruits).

Chemical Constituents

Protein, fats, carbohydrates, minerals (Ca, P, Na, K, Fe), thiamine, riboflavin, niacin, vitamin A and C, volatile oil (keton, carvacrol), di-limonene (formerly called carvone)

Processed Products

The two important processed products from caraway seed are: (i) volatile oil and (ii) carvone, isolated from oil of caraway. They are briefly discussed as under:

1. *Volatile Oil*: The volatile oil contains a mixture of ketone, carvone, a terpene, formerly called carvone, but now

recognized to be dl-limonene and traces of carvacrol. Pure carvone is prepared by decomposing crystalline compound of carvone with hydrogen-sulphide.

2. *Decarvonized Oil*: It contains limonene with traces of carvone and is sold in the market as light oil of caraway. It finds use in scenting cheap soaps.

3. *Caraway Chaff Oil*: Caraway chaff oil is said to be distilled from the husks and stalks that remain after threshing. The dried exhausted and pulverized caraway chaff contains crude protein and fat. It can be used as a cattle feed.

Properties and Uses

Caraway is widely used for culinary purposes as a spice and for flavouring rye bread, biscuits, cakes and chesse. It is also used in manufacturing 'Kummel' cordial. It is used as an ingredient for seasoning sausages and pickling spice, soups and meat stews. It is also used in scenting soaps.

In Medicines

It is a mild stomachic and carminative. It is occasionally used in flatulent colic and as an adjuvant or corrective for medicines. Its volatile oil is generally employed.

Its content Carvone is used as anthelmentic in hookworm disease. Caraway oil is used chiefly for flavouring purposes and in medicine as a carminative. It is also used to correct the nauseating and groping effects of medicines. It is recommended for scabies along with castor oil. The caraway oil and alcohol is beneficial in scabies. Seed oil is used orally in overcoming bad breath and insipid taste.

12. Celeriac
Apium graveolens var. *Rapaccum* de Candole

Apiaceae (Umbelliferae)

Celeriac is another cultivated variety of *A. graveolens* somewhat smaller in size than celery. It has dark-green foliage with less developed stalks and swollen roots. This is also called turnip-rooted celery. It is grown to a limited extent in India (Uttar Pradesh and Punjab) and in France.

Useful Parts

Roots, leaves, stalks.

Chemical Constituents

The leaves of celeriac contain carbohydrates, fat, protein, minerals, like calcium, phosphorus; iron; vit. A, Vit. C. The stalks contain, carbohydrates, fat, protein, calcium, phosphorus, iron, vit. C. The herb is also reported to contain glucoside apin.

Properties and Uses

The leaves and stalks are used as salad and for flavouring soups. The tuberous root is consumed after cooking. It is diuretic and is given in colic.

13. Celery Seed
Apium gravelolens Linn.var. Celery
Apiaceae (Umbelliferae)

Celery seeds have been known for over 3000 years. It was mentioned in Homer,s Odyssey, written about in the 17th century BC, as an excellent medicament. The earliest recorded use of celery seeds as a sesoning for food was not until 1623 in France. Celery is grown for its leaves, seeds, oleoresin and essential oil. The native habitat of celery extends from Sweden to Egypt, Algeria and Ethiopia and in Asia, to India, Caucasus and Baluchistan. Celery seed is also grown in France and China. The Chinese celery seed, mainly grown in Gaungdang Province near Hong Kong and in Gansu in the North West China, is smaller and milder than the Indian celery seed and has more striations. It is also claimed to be the first mentioned as a cultivated food plant in France in 1623. The plant has a grooved, fleshy erect stalk, a tap root, radical and stalked leaves, hermaphrodite flowers and humped ovoid light brown aromatic seeds up to 2 mm in diameter. The rigid fruit is small ovoid, 1.1 to 1.5 mm long an 1-2 mm in diameter, contains a small brown seed, with united or separated pericarp. The epicarp is intercepted with oil ducts.

Useful Parts

Leaves, stalks, seeds

Chemical Constituents

Leaves: Proteins, fats, carbohydrates, minerals (Ca, P, Fe), Carotene, riboflavin, niacine, vitamin C.

Seeds: Proteins, fats, carbohydrates, minerals (Ca, P, Fe, Na, K), thiamine, riboflavin, niacine, vitamin A and C.

Properties and Uses

The dried ripe fruits (seeds) are used as a spice. Leaves and stalks are used as salads and in soups as a pre-dinner appetizer. The dried ripe seeds are used as a spice to flavour soups, salad, tomato juice and sauces. The celery seed or one of its extractives in soluble form are used extensively in many meat seasonings, in flavouring non-alcoholic beverages, confections, chewing gums, ice-creams and baked goods. The commercial use of celery seed oil is as fixative and as an important ingredient in novel perfumes. Celery seed oil is one of the most valuable flavouring agents being a warm, aromatic and has pleasing flavour.

In Medicine

The seeds are used for ailments like rheumatism, arthritis and nervousness, while some expert herbalists prescribe the plant for every ailment from rheumatism and sciatic to liver disorders, hypertension and even poor eyesight. The seeds are not only emenagogue, a substance that induces menstruation in non-pregnant women, but are also condemned for attempted use as an illegal abortifacient. In Egypt, wild celery seed is used as a diuretic, digestive aid, emmenagogue. It is also recommended as an aphrodisiac and anti-lactogen for nursing mothers. In the indigenous system of medicine, the oil is used as a nerve tonic and in digestive mixtures.

Seed are stimulant and tonic, and administered in asthma and for liver disease. Essential oil contracts graved and virginal uterus. As adomestic medicine, seeds are used as a nervine sedative and tonic. A domestic remedy for rheumatism is 1-in-20 dilution. Celery seeds are the main ingredient of celery tonics. The fruits also yield 7 per cent of a fatty oil-'oil of celery'. This is used as an anti-spasmodic and nerve stimulant. It has been successfully employed in rheumatoid arthritis and probably acts as an intestinal antiseptic.

It is considered diuretic and is given in anasarca and colic. Celery seed serves as a 'bird feed' also.

Celery seeds relieve flatulence, increase the secretion and discharge of urine. They also promote sexual desire. They are tonic, laxative and stimulant. They are beneficial in counteracting spasmodic disorders. The alkaline elements in the celery is effective in diseases resulting from acidity and toxemia. Seed extract is helpful in gout and rheumatism. Seeds are likewise are very beneficial in nervous afflictions and disorders of respiratory system *e.g.* asthma, bronchitis, pleurisy and tuberculosis of lungs. Dry root powder is a valuable tonic in general weakness and malnutrition. Leaf juice with honey is useful in sleeplessness.

Processed Products

Seed-based Processed Products

1. *Celery Seed Oil*: The aroma of celery seed oil is warm, spicy, slightly fatty, fruity, penetrating and very persistent. It has a burning sensation and is very bitter.

   The celery seeds on steam distillation yield 2-3 per cent of a pale-yellow volatile oil with a presistent odour, characteristic of the plant. In trade, this is known as 'celery seed oil' and is much valued both as a fixative and as an ingredient of novel perfumes, pharmaceutical preparations and medicines.

2. *Celery Herb (leaf) Oil*: The volatile oil content of celery leaves is too low to permit their economic distillation and hence it has not attained any commerical importance at present. The celery herb oil is quite different from seed oil, the former being more representative of the plant as such. This new type of herb oil could be introduced to a great advantage in many products of our food industry, notably soups to impart fresh celery leaf flavour with advantage.

3. *Celery Chaff Oil*: It has rather harsh and coare odourand flavour than that of celery seed oil. It is frequently adulterated with celery chaff oil or with d-limonene, the addition of which is difficult to detect. The odour and flavour of Indian celery seed oil is normal, but usually not quite as fine as those of French celery seed oil.

4. *Celery Seed Oleoresin*: It is green having a volatile oil content. The oleoresin of celery seed is of two types: French and Indian. The former is sweet, herbal and tenacious with only a slight citrus undertone. The latter type is more herbal with a slight lemon-like aroma and tenacious herbal undertones. The French product is more pleasing; however, both are quite bitter. The Oleoresin of celery seed is prepared by extraction of crushed dried celery seeds with suitable volatile solvents like food grade hexane or

ethylene dichloride, filtration and desolventization under vaccum. The oleoresin not only possesses the volatile top note of the celery essential oil, but also the 'body' *i.e.* the fixed or non-volatile extractive matters of the celery seed.

Celery seed oleoresin is More convenient to use and store, help maintain cleanliness, easy to store, imparts uniform standard flavour to the cold product, microbial and fungal contamintion eliminated, no enzmatic problem, and no fumigation is necessary as is necessary during storage of spices/sses. Colour distribution also of uniform shade on proper mixing and more stable than whole or ground spices.

5. *Celery Salt*: It is prepared by mixing finely ground table salt with ground celery seed or celery seed oleoresin or ground dried celery stems.

6. *Celery Pepper*: It substitutes ground black pepper for celery salt. It should contain not more than 70 per cent of pepper.

7. *Dehydrated Celery*: It is commercially available as: (a) 10 mm celery stalk dice, (b) leaf and stalk flakes, (c) stalk and leaf granules, and (d) celery powder. These are used to flavour soups, for broth base, for canned tuna fish, stuffings and stewed tomatoes and as a garnish on potato salad and meat sauces.

8. *Celery Stalk Products*: The stalk products exclusively have the deepest green colur. It is frequently protected by the addition of minute amounts of sodium bisulphite or sodium sulphite. Celery flakes are used in dry soup mixes, canned soups, sauces, stuffings, casserol products and vegetable specialities. Granulated or powdered celery is a good choice for canned and frozen sauces and dry mixes for breadings and soups. Cross-cut and diced celery is used in canned and frozen soups, relishes, vegetable specialities, and salad mixes.

9. *Processed Celery Juice Blends*: (a) Celery: Tomato juice blend from fresh celery leaves and tomatoes has been successfully prepared and marketed (b) Freeze-dried Celery: Cross-cut slices of celery stalks are available in freeze-dried form. This process is effective in retaining celery's original shape

and crispness. This product makes a crisp garnish for potato salad, casserols, Chinese dishes, elatin salad, pickles and relishes.

15. Chillies (Capsicum)
Capsicum annuum Linn.

Solanaceae

Chillies are the native of America. Columus C1452 A.D.) brought chillies on his fired voyage back from America. Then chillies spread along the mediterrnen coast and in India. It was introduced in Goa by Portugucse.

The dried ripe fruits of the species of genus *Capsicum* are chillies. They are called red peppers or capsicums. They constitute an important well-known commercial crop. These are used as a condiment, culinary supplement or as a vegetable. Chilles are virtually an essential item in the kitchen. *Capsicum annuum* is a variable annul sub-shrub (originated in the American tropics). The fruits are usually pendent, which provide us all red pepper. A mild form with large inflated fruits of sweet pepper cayenne, paprika and chillies used as a vegetable. Red peppers or cillies are cultivated mainly in tropical and sub-tropical countries *e.g.* Africa, India, Japan, Mexico, Turkey and the USA.

Nearly all varieties of low and medium pungency that are cultivated on field scale in India belong to *Capsicum annuum*. There are only a few perennial chilli varieties which are characterized by small size of pods and high pungency and are rarely cultivated on field scale, such as 'Bird chilli' and 'Tabasco chilli' belonging to *Capsicum frutescens*. Chillies known to Indians about 400 years ago, only when this crop was first introduced into India by Portuguese during the last quarter as the 15th century. Their cultivation became popular in the 17 the century. They are a native of South America and their cultivation was known to natives of Peru since prehistoric times.

Chillies constitute the most important spice grown all over the world except in colder parts. There are many varieties which differ in habit, size, shape, colour and pungency. Commercially, chillies may be classified on the basis of their colour, size, pungency and the end use to which they are put. There are four species in cultivation throughout the world. However the species, *viz., C. pendulum* and *C. pubescens* are mostly confined to South and Central America. The other two species, *viz., C. annuum* and *C. frutescens* are cultivated throughout the world. The former is the most commonly cultivated species. The chillies grown in India mostly belong to *C. annuum*.

The chillies are broadly divided into two varietal groups: (1) the long, pungent type, including pickling type, used as a spice and (2) bell-shaped, non-pungent or mild and thick-fleshed type, popularly known as 'Shimla Mirch' or 'Paprika' which is commonly used as a curried-vegetable. 'Paprika' also belongs to mild group of *C. annuum*. The varieties vary in colour, from red to yelow or white, in shape from long and thin to round or oblong, and in lenght from one cm to thirty cm. Good fruit length, shining red colour, high pungency and strong attachment of calyx are important quality factors.

Useful Parts
Fruits.

Chemical Constituents
Protein, fats, carbohydrates, minerals (calcium, phosphorus, iron), oleoresin, fixed (fatty) oil, volatile oil, capsaicin (capsicutin), carotene, thiamine, riboflavin, niacin, vitamin-C.

Processed Products
1. *Dehydrated Green Chillies*: There are standard conditions and pretreatments required to produce best quality dehydrated green chillies.
2. *Drying Yield of Chillies*: Dried chillies generally contain about 6 per cent stalks, 40 per cent pericarp and 54 per cent seeds. Important constituents of colour and capsaicin are concentrated in pericarp. About 90 per cent of the capsaicin in chillies has been noticed in placenta. Placenta which represents only less than 4 per cent of total weight has a capsaicin content of about 7 per cent. In

trade, the dried forms of fruits of *Capsicum* species are categorised as: (i) highly pungent 'chillies', (ii) moderately to mildly pungent 'capsicums' and (iii) 'paprika' which may be sweet or mildly pungent. Paprika is always ground product, and chillies and capsicums are in trade in whole or ground form. All above three types are also extracted with solvents to prepare their oleoresins. Blends of ground chillies and capsicum are available in market as cayene and red pepper or mixed with other spices for the preparation of 'chilli powder'. Other products include larger-fruited, sweet or mildly pungent varieties of *C. annuum* used in fresh state as vegetable or in preserves.

3. *Types of Oleoresins*: There are three different types of oleoresins obtained form dried fruits of *Capsicum* species. There are supplied in free-flowing from or dispersed on suitable carriers.

   (a) *Oleoresin Capsicum (Bird Chillies i.e. C. frutescens)*: Oleoresin is prepared from the most pungent, small-fruited bird chilles. This oleoresin has a very high pungency. It is used exclusively for official pharmaceutical work. It is useful to impart pungency to manufactured foods and some beverages. It is evaluated solely on its content of capsaicin. Commercial capsicum oleoresins are usually supplied in pungency ratings between 500,000 and 1,800,000 Scoville units (approximately 3.9-14 per cent capsaicin w/w) with colour values expressed on the ASTA (American Spice Trade Association) scale of 3,500 units maximum and 400 units maximum in decolourized types.

   (b) *Oleoresin Red Pepper (C. annuum):* It is prepared from the longer, moderately pungent capsicums which are used in the production of red pepper. Commercial red pepper oleoresins are usually supplied in pungency ratings ranging between 80,000 and 500,00 Scoville units (approximately 0.6-3 per cent capsaicin w/w) and in wide colour range of up to a maximum of 20,000 colour units.

   (c) *Oleoresin Paprika:* It is obtained from varieties of *C. annuum* from which paprika is produced. It has a high colour value, but little or no pungency. Commercial paprika oleoresins are available in colour strengths ranging from 12,000 to 100,000 units; the 40,000-80,000 range being the most popular.

4. *Fixed Chilli Seed Oil*: The fixed (fatty) oil in fruit is distributed unevenly, being mainly found in seed. As with some other constituents of the fruit, the fixed-oil content gradually increases during maturation from green to the ripe red stage.

   Analyses of the fixed-oil composition have been concerned almost exclusively with paprika and similar large form of *C. annuum*. The fixed oil has been found to comprise tryglycerids (about 60 per cent) in which linoleic and other unsaturated acids predominate. Only a small number of samples have been analysed and some differences between cultivars are apparent. It seems that seed fat and pericarp fat of paprika types are distinguishable. The fat content and composition of paprika powder, and its propensity to autoxidation and perhaps also to discolouration are, therefore, dependent upon whether seeds are removed from the pods before grinding.

5. *Volatile Oil*: The fruits of *Capsicum* species have relatively low volatile-oil content. This has been reported to range from about 0.1 to 2.6 per cent in paprika and similar large forms of *C. annuum*. The initial volatile-oil content of freshly picked fruit is dependent largely upon the species and cultivar grown and the stage of maturity at harvest. The eventual volatile-oil content of dried product, however, may be lower. It is dependent upon drying procedure, the duration and condition (whole or ground) of storage. Paprika powder, for example, usually contains less than 0.5 per cent of volatile oil.

Properties and Uses

As a spice in all types of curried dishes in India and abroad dry chilli is widely used. Curry powder is made by grinding roasted dry chilli with other condiments such as coriander, cumin, turmeric and farinaceous matter. It is used as well for seasoning of egg, fish and meat preparations, sauces, chutneys, pickles, frankfurters, sausages etc. 'Mandram' is a West Indies stomachic preparation made by adding cucumber, shallot, lime juice and Madeira wine to mashed fruits of bird chilli. Paprika, red pepper, Kashmiri mirch, degi or degchi mirch which are mild in pungency, are used to colour, flavour and garnish dishes. Shimla mirch is not pungent and used as curried vegetable. It is specially rich in vitamin C. One Shimla mirch a day fulfils our daily requirement of vitamin C.

In Medicine

Some preparations are used as counter-irritants in lumbago, neuralgia and rheumatic disorders. Capsicum has a tonic and carminative action and is specially useful in atonic dyspepsia. It is sometimes added to rose gargles for pharyngitis and it relaxes sore throat. It is administered in the form of powder, tincture, linament, plaster, ointment, medicated wool, etc. In some, it is administrered in the form of powder, tincture, linament, plaster, ointment, medicated wool, etc. In some of the preparatins, 'Oleoresins Capsici B.P.C. syn. Capsaicin, 'the alcohol-soluble fraction of ether extract of capsicum is the active ingredient. The common green chillies (*Capsicum annuum*) are the most fertile source for the enzyme, L-asparaginase which has an anti-tumour element and is used in the treatment of 'Acute Lymphocytic Leukaemia' a type of cancer.

Apart from the use in food processing industries, pharmaceutical and cosmetic industries use chilli oleoresin of high pungency and low colour. This oleoresin known as 'oleoresin capsicum' is used in pain-balms, vaporubs, linaments, etc, since the pungent principles 'capsaicin' serves as an effective 'counter-irritant'. Seed oil is a byproduct and can be used for edible purposes The solvent-extracted residue from chilli seed can be used as a fertilizer or as an animal feed.

15. Clar Sage (Garden Sage) *Salvia sclarea* Linn.

Lamiaceae (Labiatae)

Clary Sage is an exotic plant and was introduced into India long back as an ornamental plant in the gardens. This is commonly known as 'Garden Sage' or Clary Sage. It is also introduced at Solan (India) from Bulgaria and has established well under mid-hills and low valley areas of Himachal Pradesh.

Its leaves are large, broad and hairy leaves. It bears a long flowering shoot with light purple flowers. The flowering shoots on steam distillation yield a commercially important essential oil with

typical fragrance which blends well with lavender, bergamot and Jasmine oils.

Useful Parts

Entire herb, leaves.

Properties and Uses

It has also found a place in cosmetics, flavouring of alcoholic and non-alcoholic beverages.

16. Clove *Eugenia caryophyllus* (Spregal) Bullok *et* Harrison

Syn: *Syzygium aromaticum* (L.) Merr. and Perry
Myrtaceae

The term 'clove' is derived from the French word 'clove' or 'clou' and the English word 'clove', both meaning 'nail'. The flower bud of

the clove tree is divisible into an elongated and a globose part simulating a nail. Clove is one of the most ancient spices of the Orient, known since 100 B.C. It was later known to the Chinese in 266 B.C. Clove was imported into Europe in 1265. Its source and place of origin both were shrouded in mystery.

The clove of commerce is the air-dried unopened flower-buds obtained from a handsome, medium-sized, evergreen, straight-trunked tree that grows in Kerala and Tamil Nadu. The tree grows to a height of 10.7-12 m. It begins flowering in about seven years and continues to produce for another 80 or more ears!

The Portuguese discovered the Moluccca Island of Indonesia in the 16th century. The French introduced the clove tree in Mauritius and Reunion in 1770. They reached to Zanzibar. Clove was established in Ceylon in 1796 before the arrival of the

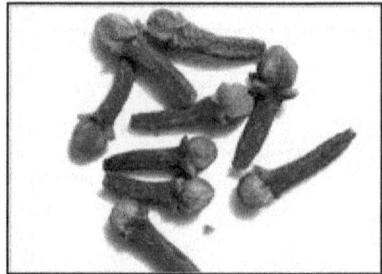

British. For sometime, clove was a Portuguese monopoly. Later, it was monopolized by Dutch. In India, clove was introduced in 1800 A.D. before the East India Company. By far the biggest clove-producing region in the world today is Zanzibar, followed by Pemba Madagascar and Indonesia. Clove is also produced in Malasia, Srilanka and Haiti.

Useful Parts
Flower buds.

Chemical Constituents
Dry clove: protein, fat, carbohdrates, minerals (Ca, P, Fe) thiamine, riboflavin, essential oil (eugenol), eugenol acetate, caryophyllene.

Processed Products
1. *Clove-bud Oil*: The volatile oil is obtained from the dried buds by steam distillation. It contains as its principal constituents, free eugenol, eugenol acetate and carophyllene. There are not responsible for the characteristic fresh and almost furity note of the pure clove-bud oil contains a substantial percentage ofeugenolacetate, whereas clove-stem oil and clove-leaf oil contain only traces of it.

2. *Clove-stem Oil*: The chief constituents, present in clove-bud oil occur also in the stem oil, but in somewhat different ratios. The percentage of free eugenol present in the stem oil is usually somewhat higher than that present in the bud oil. The stem oil, on the contrary, contains only a small amount of eugenol acetate, whereas the bud oil has been reported to contain up to 17 per cent of this ester. Substances imparting the characteristic, almost fruity odour to the bud oil, seem to occur in the stem oil in still lesser quantities, or lack entirely. This explains the coarser and 'flatter' odour of the stem oil. Clove-stem oil contains a few constituents which have not been noticed in clove-bud oil.

3. *Clove-leaf Oil*: Clove-leaf oil generally contains a lower percentage of total eugenol to that present in clove-bud oil, and eugenol acetate occurs in the leaf oil as in the stem oil, however in very small quantities, The trace substance

methyl-n-amyl ketone for example, which imparts characteristic, almost fruity odour to the bud oil, occurs in the leaf oil, probably in even more minute quantities than in the stem oil. As far as the substances which occur in the stem oil (but not in the oil) are concerned, *viz.* a sesquiterpene alcohol and naphthalene. Only the latter (traces) has been noted in clove-leaf oil.

4. *Oil of Mother Cloves*: The oil distilled from mother cloves (yield:6.5 per cent) has phenol content: a solid phenol, having paraffin-like odour.

Clove-root Oil

It is obtained by steam distillation of the root. It is bright yellow when freshly distilled. It compaers with the oil from clove-buds in composition, odour and quality. It contains 85-95 per cent eugenol.

Properties and Uses

In all Indian homes, it is used as a culinary spice, as the flavour blends well with both sweet and savoury dishes. Cloves, both whole and ground, are used in baked goods, cakes, confectionery, chocolate, puddings, desserts, sweet syrups, preserves, etc. Cloves are used for flavouring curries, gravies, pickles, ketchups, sauces, spice mixtures and as pickling spice. Clove has stimulating properties and is one of the ingredients of betelnut chew. In Java, clove is used in the preparation of a special brand of cigarette for smoking. It is an ingredient of many tooth pastes and mouth washes. The oil has many industrial applications. It is extensively employed in perfumes. It is extracted and used as an imitation carnation in perfumes, and for the formation of artificial vanilla. The use of clove in toilet waters, and soaps of oriental and spice odour is well known.

In Medicine

It is highly valued in medicine as carminative, aromatic and stimulant. It is also used in flatulence and dyspepsia. The essential oil has many more uses. It is used in medicine as an aid to digestion and for its antiseptic and antibiotic properties in tooth-ache. Externally, it has a counter-irritant action.

Because of its antiseptic and anti-bacterial properties, numerous pharmaceutical preparations contain oil of clove. It has also been reported to help diabetics in sugar assimilation.

It helps simulate sluggish circulation and thereby promote youthful digestion and metabolism. It benefits in cramps, painful conditions of joints, neuralin and migraine. It is useful in rheumatic application.

17. Coriander
Coriandrum sativum Linn.

Apiaceae (Umbelliferae)

The coriander is native to mediterranean Europe and Western Asia. Coriander has been used by man since long as its seeds were found in the Egyptian tombs of 960 BC. Further, Persia grew coriander 3,000 years ago and it added fragrance to the Hanging Gardens of Babylon. Hebrews knew it also, since in the Old Testament in the Book of Exodus, 'manna' is described as being like white coriander seed. In the 3rd century BC, Romans also found coriander seed as an

excellent seasoning for popular foods. Hippocrates and other Greek physiciations employed it in their medicines.

Useful Parts

Leaves, Fruits.

Chemical Constituents

Volatile oil (terpene-d-pinene, coriandrol), essential oil, fatty oil (palmitic, oleic, linoleic and patrosilinic acid), oleoresin, minerals (Ca, P, Fe), carotene, thiamine, riboflavin, and niacine.

Processed Products

1. *Volatile Oil*: The volatile oil is made up of hydrocarbons and oxygenated compounds. The hydrocarbons account for about 20 per cent of the essential oil. The major oxygenated compounds present are d-linalool or coriandrol (45-70 per cent). The oil causes irritation when it is in contact with skin for a long time. The unripe fruits and other parts of the plant produce an inferior oil with a bug-like odour, which on keeping disappears evidently due to the polymerization of the odoriferous principles.

2. *Fatty Oil (Non-Volatile Oil)*: The seeds also contain 19-21 per cent of a non-volatile fixed fatty oil having a dark, brownish-green colour and an odour similar to that of coriander oil. The components of insoluble fatty acids are: palmitic, patrosslinic, oleic and linoleic. The oil solidifies on keeping. It is soft in consistency and green in colour.

3. *Coriander Herb Oil*: Volatile oil is obtained from stalks and leaves. It is absolutely impossible to use the leaf oil instead of seed oil in any flavour, perfume or pharmaceutical formulae. The leaf or herb volatile oil has a pronounced odour of decylaldehyde or other higher fatty aldehyde.

4. *Coriander Oleoresin*: Recently oleoresin has been prepared. Coriander has more fatty oil than volatile oil, and the former interferes in the extraction of the oleoresin. It is desirable to lower the fatty oil and enhance volatile oil suitably. Volatile oil content of commercial oleoresins ranges from 5 to 40 per cent.

5. *Dhania Dal*: This is mainly used in Gujarat, Maharashtra and some Northern States as an adjunct in superior

exclusively. The seed is dehusked and flanked and then given a mild heat treatment and salted. The treated seeds may also be eaten as a highly flavoured digestive chew.

6. *Soluble Coriander (Superesin)*: This is prepared by properly blending and dispersing a minimum of 3 per cent of total extractives of coriander on a soluble, dry edible carrier. The coriander extractives are the sum of the non-volatile ether extract and volatile oil content. The coriander flavouring consists of a maximum of 33 per cent of volatile oil of coriander and the remainder as the non-volatile ether extract of coriander. The coriander extractives are of good quality and possess the characteristics flavour of good grade coriander. Their flavour is unimpaired in or during the process of extraction.

7. *Seasoning from Coriander Roots*: In Thailand, they derive yet a third 'seasoning' from coriander roots which are claimed to be as distinctly different in flavour from seeds, as leaves are from seeds.

In the USA, different products of coriander are in the market: (a) Dried and freeze-dried green leaves of coriander (b) Dried coriander seeds (whole and ground) (c) Coriander essential oil (d) Coriander oleoresin (e) Thai 'Coriander Root Seasoning' for foods.

Properties and Uses

The fresh green stems, leaves and fruits of coriander possess characteristic and pleasant aromatic odour. The entire plant, when young, is used in preparing chutneys and sauces, and the leaves are used to flavour and garnish curries and soups. The fruits (seeds) are widely employed as condiment with or

without roasting in the preparation of curry powders, pickling spices, sausages and seasonings. They are used for flavouring pastry cookies, buns, cakes and tobacco products. In the USA and Europe, coriander is emploed for flavouing liquors particularly gin. Coriander seeds are generally used after mild roasting. Coriander is one of the important ingredients in the manufacture of the food flavourings *e.g.* (i) bakery products, (ii) imitation flavour, (iii) pork, frankfurter, meat, fish and salads, (iv) soda and syrups, (v) gelatin dessert and puddings and (vi) candy preserves, (vii) liquors. It is used by soft drink manufacturers, pickles and sausages makers, and cigarette and cosmetic manufacturers. Oil of coriander seeds is soft, pleasant, slightly spicy note blends into scents of oriental character. It harmonizes well with jasmine, imparting life and lift to otherwise dull or too synthetic compositions. Decyladehyde is also useful in perfumery. Decyladehyde is obtained by treating the volatile oil and bisulphite. The volatile oil is used to flavour cocoa and chocolate. The oleoresin is used for flavouring beverages, pickles, sweets, sausages and other delicacies, snacks, etc. The residue from distillation can be used as a good cattle feed, as it is rich in protein and fat.

In Medicine

Coriander seeds are carminative, diuretic, tonic, stomachic, antibilious, refrigerant and aphrodisiac. They are used chiefly to conceal the odour of other medicines and to correct the griping qualities of rhubarb and senna. The seeds are chewed to correct foul breath. They are also considered to lessen the intoxicating effects of spirituous liquors. An infusin of seeds in combination with cardamom and caraway seeds is useful in flatulence, indigestion, vomiting and intestinal disorders. Leaves relives flatulence, increase secretion, discharge urine and reduce fever. The juice of fresh leaves is beneficial in indigestion, piles, dysentery, hepatitis, ulcerative colitis and in typhoid fever. Seeds reduce feverishness and promote feeling of coolness. Coriander juice is highly beneficial in deficiencies of vitamins A, B, B_2, C and iron. Decoction of seeds is helpful to treat dyspesia, flatulent colic and biliousness. Dry seeds are powerful cholesterol lowering food. Seeds are beneficial in checking excessive menstruation. Seeds are powerful aphrodisiac as well. Juice from fresh leaves alongwith turmeric powder is effective in case of pimples, dry skin and black heads. Decoction of seeds alongwith milk, jaggery or honey is valuable in piles. Poultice of seeds cures headache.

18. Cumin Black
Nigella sativa Linn.

Ranunculaceae

Black cumin is the dried seed-like fruit. It is a small herb, about 45 cm in height, and native of the Levant (Eastern Mediterranean region). It is said to be cultivated on a small scale. Black cumin (*Kalonji*)-growing districts are Rewa, Sidhi, Mandla and Bilaspur of Madha Pradesh.

Useful Parts

Seeds (fruits).

Chemical Constituents

Volatile Oil (carvone, d-limonene, cymene, nigellone), fixed oil (linoleic aicd, oleic acid, palmitic acid), nigellin, tannins, resins, saponins, amino acids (cystine, lysine, aspartic acid, glutamic acid, alanine, tryptophan, valine and leucine).

Processed Products

1. *Volatile Oil*: The seeds yield, on steam distillation, a yellowish-brown volatile oil with an unpleasant odour. The oil contains carvone (45-60 per cent) d-limonene and cymene, a carbonyl compound, nigellone.

2. *Fixed Oil*: The fatty oil obtained by the expression of seeds is reported to be used for edible purposes. Extraction with benzene and subsequent steam-distillation of extract to remove volatile oil gave about 31 per cent of reddish-brown, semi-drying oil. Besides the volatile and fatty oils, black cumin seeds contain a bitter principle (nigelline), tannis, resins, proteins, reducing sugars (mostly glucose), saponins and arabic acids. The free amino acids present in dormant seeds are *viz.*, cystine, lysine, aspartic acid, glutamic acid, alanine, tryptophan, valine and leucine.

Properties and Uses

It is used as a spice extensively in India in various culinary preparations. It is also used to flavour foods and beverages. Seeds are also scattered between folds of linen or woolen cloths to preserve them against insect attack.

In Medicine

The seeds are carminative, stimulant, diuretic, emenagogue, and galactogogue. They are used in the treatment of mild cases of puerperal fever. They are externally applied for skin eruptions. They are also used against scorpion sting. The volatile oil has a carbonyl compound 'nigellone'. It protects guinea-pigs against histamine induced brancho-spasm.

19. Cumin Seed (*Safaid Jeera*)
Cuminum cyminum Linn.

Apiaceae (Umbelliferae)

Cumin is one of the oldest spices, known since Bilbical times. The plant is grown extensively in Iran, India, Morocco, China, Southern Russia, Southern Europe, UAR and Turkey. Excavations in the Indus Valley revealed that cumin, fennel, fenugreek, coriander, garlic and mustard had been used even before about 1000 BC. These spices were included in the ancient Susruta, Mushkakadigana or 'herbal medicines' prescribed to remove fat and to cure urinary complaints, piles and jaundice. Cumin seed is one of the ancient spices, mentioned a number of times in the Old Testament. It was once called the best of all condiments by the Roman naturalist, Pliny. Its strong, distinctive flavour was much used as a seasoning and also as a medicine. It was used by Egyptians in 5000 BC to season meats, fish, stars and to mummify their dead. Cumin (safaid zeera) comprises the dried pale-yellowish to light greyish-brown seeds of a small slender annual herb of a native of Egypt and Syria, Turkestan and the Eastern Mediterranean region. It grows to a height of 30-45 cm. The aromatic fruit 'seed', is elongated, oval-elliptical, deeply furowed, approximately 6-mm long, and light yellowish-brown, somewhat similar to caraway seed, but slightly longer. The odour is peculiar, strong and heavy flavour is warm, slightly bitter and somewhat disagreeable.

The Romans were said to have substituted it for pepper, making it into a paste which was then spread on to bread. By Middle Ages, cumin seed had become one of the most popular spices in Europe and Britain. The well-known types of cumin in the international trade are:

1. *Iranian Cumin*: The Iranian cumin seed is pale-green and has 2.5 to 3.5 per cent essential oil content. It is a 'machine cleaned product' which is marketed simply as Iranian cumin without grade distinctions. Iran also has a 'wild herb' which the natives call 'black cumin'. It has similar flavour, though sweeter than *Cuminum cyminum*.

2. *Indian Cumin*: The flavor and aroma characteristics of Indian and Iranian cumines are quite similar. The Indian product is golden-brown and essential oil contents of 'prime quality' seed ranges from 3 to 5 per cent and richer than the Iranian cumine in oil content.

3. *Egyptian, Turkish Cumin and Cumin from other Origins*: Egypt and Turkey are fairly constant exporters to a small degree. Cumin from these and other sources can vary from 1.5 to 5 per cent in essential oil content. The flavour character is apt to be different from either Indian or Iranian cumin seeds. It is somewhat inferior.

Useful Parts

Fruits.

Chemical Constituents

Protein, fats, carbohydrates, mineral (calcium, phosphorus, iron, sodium: potassium), vitamins, vit. B_1, vit. B_2, vit. C, vit. A, niacin, volatile oil (cumin aldehyde, cuminol, carvone, cynol, terpene), essential oil, oleoresin.

Processed Products

1. *Cumin Powder*: It is a good value-added product and like oleoresin and essential oil, it seems to have a good future in world trade.

2. *Volatile Oil*: The dried fruits are steam-distilled to yield 2.5 to 4.5 per cent of valuable volatile oil, colourless or pale yellow, turning dark on keeping. The older seeds contain less oil. The chief constituent of the oil is cuminaldehyde, which is used in perfumery.

3. *New Super-Critical Fluid Extraction (SCFE) Process for Volatile Oil*: A new extraction process for isolation of essential oil/ extractives from natural substances using liquid and dense carbon-dioxide has been described. Cardamom, cumin, clove, ginger, parsley, mace, sandalwood and vetiver are used as extraction materials. The essential oil/extractives obtained by this method are found superior in quality and flavour as compared with conventional steam-distilled essential oil.

4. *Fixed Oil*: The seed also contains about 10 per cent fixed (non-volatile), greenish-brown oil with a strong aromatic flavour. It is a semi-drying oil having an iodin value of 92.

5. *Cumin Oleoresin*: Cumin seed oleoresin is also manufactured.

Properties and Uses

Cumin Seeds are largely used as a condiment and form an essential ingredient in all mixed spices and curry powders for flavouring soups, pickles and for seasoning breads and cakes. They are also candied. Cumin is used to season some kinds of cheese in Holland and Switzerland. It is used to flavour bread, cakes and pastries. Cumin makes an excellent seasoning for soups and stews, for which it is widely used throughout Latin America. It is also employed in native dishes of Central and South America. Most of the cumin seed used in the USA goes into chilli powder, followed by manufacturing 'chilli concame' and 'chilli meat products', and then other Mexican-style foods It is also an excellent spice for rice dishes, curries, stuffings, sauces and marinades, particularly for Middle Eastern dishes. After chilli, peppers, cumin is perhaps the most distinguishing flavour characteristic of Latin American foods in general. The volatile cumin seed oil is employed advantageously in flavouring especially in curries and culinary preparations of oriental character. The oil is used in perfumery and for flavouring liqueurs and cordials. The residue left after the extraction of volatile oil contains protein and aromatic greenish fixed oil or fats. Its cake can be used as a cattle feed. The fixed oil could also find use in the oil, fats and soap industries.

In Medicine

Cumin seeds have long been considered stimulant, carminative, stomachic, astringent and useful in diarrhea and dyspepsia. They are now chiefly used in veterinary medicine. An infusion of cumin relieves from flatulence. Digestive disorders such as biliousness, morning sickness, indigestion, colic and flatulence are cured using infusion of seeds. Seed powder is beneficial in the treatments of jaundice, nausea and vomiting. Chewing seeds after food, prevents dental caries, indigestion and constipation.

Seeds are calming and sedative. They soothe the nervous system and thus benefit in sleeplessness. Decoction of seeds is an antiseptic beverage and useful in common colds and fever. It also soothes irritation of throat. Cumin seeds, fried and ground, consumed with honey helps in the treatment of sinusitis. Decoction of seeds is advised during the period of pregnancy. It helps the development of the body and eases childbirth. Infusion of seeds is beneficial to treat boils, bleeding from nose and lungs.

20. Curry Leaf
Murraya koenigi Spreng

Rutaceae

Curry leaves (Curry Patta or Kary Patta) have been extensively used in Indian cookery for centuries. It constitutes one of the major leafy spices of India, as an adjunct to several Indian curried dishes.

Leaves are obtained from small shrub (0.9 m) or a small tree up to 5 m in height. It is found almost throughout India and the Andaman Islands up to an altitude of 1,500 m. It is commonly found in forests, often as gregarious undergrowth, along the foot of the Himalayas from the river Ravi to Sikkim and Assam. It is also found grown in Bengal, Madhya Pradesh and in the South and South-Western States, namely Maharashtra, Tamil Nadu, Kerala, Karnataka

and Andhra Pradesh. It is much cultivated for its aromatic leaves and for ornamental value throughout India.

Useful Parts

Leaves.

Chemical Constituents

Protein, fats, carbohydrates, mineral, calcium, phosphorus, iron (as vit. A) nicotinic acid, vit. C, thiamine, riboflavin, volatile oil and oleoresin.

Free aminoacids *viz.*, asparagine, glycine, serine, aspartic acid, glutamic acid, threonine, alanine, proline, tyrosine, tryptophan, aminobutric acid, phenylalanine, leucine, isoleucine and traces of ornithine, lysine, arginine and histidine; glucoside, koenigin and a resin.

Processed Products

1. *Volatile Oil*: Fresh leaves on steam-distillation yield a volatile oil. Rectified curry-leaf oil is deep yellow in colour with a strong spicy odour and pungent clove-like taste. The major contents are 1-subinine, 1-α-pinene, dipentene, 1-terpene, 1-caryophyllene,1-cadinene.

2. *Dehydrated Curry Leaves*: Vacuum-shelf dried leaves retain a better green colour than the sun-dried or dehydrated and cross-flow or thorough-flow driers. There is complete darkening of colour in all these cases. However, the yield of volatile oil is higher both in the sun-dried and dehydrated leaves than in the vacuum-shelf dried leaves. Dried leaves yield more oil than the fresh leaves.

Propertis and Uses

The leaves of this plant have been used for centuries in South India as a natural flavourng agent in various curries, chutneys and a variety of food recipes. Fresh curry leaves on steam-distillation under pressure yield 2.6 per cent of a volatile (essential) oil. It finds use as a fixative for heavy type of soap perfume.

In Medicine

The curry leaves benefit in nausea, morning's sickness, indigestion, vomiting, diarrhoea, dysentery, diabetes, premature graying of hair, kidney disorders, etc. The leaves, the bark and the root of the plant are used as a tonic, stomachic, stimulant and carminative. An infusion of the roasted leaves is used to stop vomiting. Externally, they are also used to cure eruptions and the bites of poisonous animals. The green tender leaves are consumed for the cure of dysentery. Leaves and roots are reported to cure piles, allay heat of body, thirst, inflammation and itching. The powdered leaves are used for healing of fresh cuts and wounds. A decoction of leaves is advised with bitters as a febrifuge and the leaves have been claimed to be used with mint (*Mentha arvensis*) in the form of chutney to check vomiting. It has also been used as an antiperiodic.

21. European Dill And Indian Dill (Sowa)
Anethum graveolens L.
Anethum sowa Roxb.ex Flem.

Apiaceae (Umbellifereae)

Early Babylonian and Assyrian herbals listed it. The Romans gave their gladiators a tonic of dill in belief that it was a stimulant. The term "aneth" was earlier referred to dill. It is under this pseudonym that it appears in the Bible. Dill's most famous culinary use the 'dill pickle' is at least 400 ears old. Parkinson in his Paradisi in Sole (1629) noted, "It (dill) is also put among pickled cucumbers

where it doth very well agree, giving to the cold fruit a pretty spicy taste or relish". Evelyn (1620-1706) also praised "gherkins muriated with the seeds of dill'. Addison (1672-1719) wrote: "I am always pleased with that particular time of the year which is proper for the pickling of dill and cucumbers". By the Middle ages, dill was widely used in cooking, but an interesting anomaly was beginning to develop. This plant, which was native to the Mediterranean, was becoming a staple food additive of northern European cooking, but forgotten in the southern areas. French and Italian cook-books hardly mentioned. However, German, Scandinavian and Russian dishes made it essential. The Western World, Russia and parts of Middle East, grow the dill which is classified as *Anethum graveolens*-the so-called true dill. This is believed to be a native of the Mediterranean and Southern Russia. India and certain areas in Asia grow a type called *Anethum sowa*, or the Indian dill which is native to Northern India. The principal difference is that the sowa seeds are especially bold (large and good looking). They also tend to be longer and narrower in shape and their oil has a somewhat different profile.

Two species of *Anethum* are recognized for commercial use, and yield dill oil used in medicine, *viz.*,

1. *Anethum graveolens* (European Dill),
2. *Anethum sowa* (Indian Dill).

European dill is indigenous to Europe and is cultivated in England, Germany, Romania, Turkey, the USA and the USSR. Till recently, it was not grown in India. However, efforts have been made to acclimatize it in Jammu, New Delhi and RRL Bhubaneshwar (Orissa).

Useful Parts
Leaves, seeds.

Chemical Constituents
Essential oil, oleoresin, carvone, a-phellandrene, Vit. B, Vit. C, niacin, minerals (Ca, P, Fe, Na, K).

Properties and Uses
The green herb is used as a pot-herb and as a flavouring agent. The essential oil is also used in the manufacture of soaps, etc. The dried residue left (cake) the distillation of the essential oil from the seeds of *A. graveolens* contains - fats, and protein, It may be used as a cattle feed, after recovering fixed oil. Dill weeds (leaves) are used, both whole and ground, as a condiment is soups, salads, processed meats, susages, spicy table sauces, saurkraut and particularly in dill pickling. Dill stems and blossom heads are used for dill pickles and for flavouring soups. Ground seed is an ingredient of seasoning. Sometimes, it is used as a substitute for caraway.

In Medicine

The essential oil, dill oil or its emulsion in water, 'dill water', is considered to be an aromatic carminative, specially useful in the flatulence, colic and hiccups of infants and children. Dill helps settle the stomach, being digestive. Consumption of dill prevents constipation. Decoction of fresh leaves mixed with food prevents digestive disorders in babies. Dill oil is beneficial in hyperacidity, diarrhoea and dysentery. The Dill is a calming and sedative food. It is an ancient remedy for insomma. Liberal use of leaves by nursing mothers is useful to increase breast milk. Decoction of leaves benefits greatly in regulating menstrual flow. Leaf paste, if applied on boils, is beneficial.

Processed Products

There are different types of dill products in commerce today: (i) dill seed, (ii) dill seed powder, (iii) dill leaves (normally called dill weed), (iv) dill-seed oil, (v) dill-weed oil, and (vi) oleoresin dill. In India only seed powder and oil are traded.

1. *Dill Seed*: The seeds of *Anethum sowa* have a pleasant dill flavour, but somewhat stronger than that of dill weed. The seeds are elliptical, slightly humped and with clearly defined ridges running length-wise. The tan-colour of dill seed most distinguishable in visual point, however, is a thin, yellowish edge which frames it all around.

2. *Dill Weed (Dill Leaves)*: It is unknown why the leaves of the dill plant are to be known as 'dill weed'. The USA, primarily California, is the major producer of dill weed for the spice shelf. Egypt is another important source of supply. When the objective is dill weed for use as a spice the top 20 cm of the plant is harvested and the stems are removed. Some growers feel that it should be harvested before flowering, others say it is better afterward. In general, the leaves have a mellower, fresher flavour than seed. Dill weed has become increasingly popular in the USA in recent years.

3. *Dill-weed Oil*: It is a field-distilled product, since it is felt that the best quality comes from freshly harvested leaves. Dill has been grown commercially in the USA. As compared to dill-seed oil, the weed oil has a sharper, cleaner, fresh dill flavour and one which doesn't seems

'oily'. The dill oils are bought on carvone-content basis, but the dill-weed oil normally has less carvone than dill seed oil. This is the major dill extractive. Most of it is produced in Washington, Oregon, Idaho and Western Canada. Some also comes from Texas.

4. *Dill-seed Oil*: Small amounts are imported from Europe and India (though spice buyers here feel that most of the product that comes through Europe today is really Indian). Very little dill-seed oil is being used in the USA. The seed oil has higher carvone content which gives it a stronger and more strident flavour.

5. *Oleoresin Dill*: This is prepared from the seed. The oleoresin contains both volatile and non-volatile extractives. It is now becoming gradually popular in the world trade.

6. *Essential Oil*: On steam distillation, Indian dill yields 1.5 to 4.5 per cent volatile oil, while European dill grown under Indian conditions yields 2.5 to 4.0 per cent. The sowa herb yields 0.06 per cent of essential oil, which has a higher proportion of terpenes (aphellandrene), but no carvone. The European and American dill herb oils contain both carvone and a-phellandrene, although the carvone content (about 20 per cent) is much lower than that of the seed oil.

22. Fennel
Foeniculum vulgare Mill.

Syn. *Foeniculum officinale* All.
Apiaceae (Umbelliferae)

In Greek mythology, Prometheus concealed the fire of the sun in a hollow fennel stalk and brought it down to earth from Heaven for the human race. Pliny, an eminent Greek physician, declared that the herb enables the eye to perceive with clarity the beauty of nature. The word 'marathon' comes from the Greek for 'fennel'. The famed battle of Marathon (490 BC) was fought on the field of fennel. During the struggle, the Athenian athlete Pheidippides ran 150 miles to Sparta for aid. That is the root of our modern usage of marathon (fennel). Fennel (saunf) comprises dried, ripe fruit (seed) of cultivated varieties of *Foeniculum vulgare* Mill. of family Umbelliferae. It is a biennial or perennial aromatic, stout, glabrous herb, 1.5-1.8 m high.

It is cultivated in Mediterranean countries, in Romania and in India. It is a native of Europe and Asia Minor.

The seed is small, oblong, ellipsoidal or clindrical, 6-8 mm long, straight or slightly curved, and greenish-yellow or yellowish-brown mesocarp is deeply furrowed, 5-ridged. It possesses an agreeable, aromatic and sweet aroma resembling aniseed.

Fennel seeds vary in size, odour and taste. The major types are:

1. *Indian Fennel*: India is by far the largest producer of fennel seed. It is characterized by uniformly light or pale-green colour, plump, evenly shaped seed of generally uniform size and of rich flavour.

2. *Egyptian Fennel*: The other main source of fennel seed supply at present is Egypt. The Egyptian seeds are less uniform that the Indian in both size and shape, and the colour is somewhat mottled. Its flavour characteristics differ from other varieties.

3. *Chinese Fennel*: The People's Republic of China produces fennel seed of good quality. It is darker in colour than the Indian, and its volatile oil tests at least as high or sometimes higher.

Useful Parts

Fruits.

Chemical Constituents

Protein, fat, crude fibre, carbohydrates, mineral matter (calcium, phosphorus, iron, sodium, potassium) vitamins- vit. B_1, vit B_2, niacin, vit. C, vit. A. Volatile oil (anethole, a-phellandrene, a-limonene), oleoresin, fixed oil (palmitic, oleic, linoleic and petroselinimic acid), thiamine, riboflavin and niacin.

Processed Products

1.   *Volatile Oil*: On steam distillation of crushed fennel seed, volatile oil is obtained. The essential oil of fennel seed is a colourless pale-yellow liquid with a characteristic taste and odour. Two types of oil are recognized in commerece - sweet fennel oil from the fruits of var. *dulce* and bitter fennel oil from the fruits of var. *vulgare*. The taste and odour of sweet fennel oil are superior to those of bitter fennel oil. The main constitutent of oil from the fruits of cultivated F. *vulgare* is anethole. Oils of good quality contain 50-70 per cent anethole. The oil of sweet or Roman fennel fruits contains: anethole, a-phellandrene and a-limonene. The high percentage of anethole and the absence of fenchone are responsible for its delicate sweet odour and flavour. Terpeneless fennel oil is obtained by removing terpenes from sweet fennel oil. The volatile oil distilled from the above ground parts of wild-growing bitter fennel contains little, if any, anethole, the chief ingredient being a-phellandrene. The oil has not found any commercial application.

2. *Fixed Oil*: Fennel seeds also contain fixed oil. The component fatty acids of the oil are: palmitic, oleic, linoleic, and petroselinic acid.

Properties and Uses

The plant is pleasantly aromatic. It is used as a pot-herb. The leaves are used in fish sauce and for garnishing; leaf stalks are used in salad. Thickened leaf stalks of Florence fennel are blanched and used as a vegetable. The maximum use of fennel seed today is in Italian sausage. Fennel has gone into other sausages as well, *viz.*, Pepperoni, Cappicola and Italian leaf to name a few. In India and the neighbouring countries, they are used as a masticatory or for chewing alone or in pans. Dried fruits of fennel have a fragrant odour and a pleasant aromatic taste. They are used for flavouring soups, meat-dishes and sauces, bread-rolls, pastries and confectionery, liquors, and in the manufacture of pickles. The residue left after the distillation of essential oil from the fruit is used as a feed for cattle. It contains proteins and fats. Face pack using seeds and other ingredients rejuvenate the skin.

In Medicine

The fruits are aromatic, stimulant and carminative. They are useful in diseases of the chest, spleen and kidney. They are employed also as a corrective for less pleasant drugs, particularly senna and rhubarb. Fennel is a constituent of liquorice powder and of preparations for allaying gripping. A hot infusion of the fruits is used in indigenous medicine to increase lacteal secretion and to stimulate sweating.

The leaves are said to be have diuretic. The roots are regarded as purgative. Fennel oil is aromatic and is mildly carminative. It is useful in infantile colic and flatulence. It checks gripping in purgatives and is a good vermicide against hookworm. It is employed as a corrective for medicinal preparations with less pleasant flavour and odour and enters into the composition of 'Fennel Water' used medicinally as a vehicle for drugs. It is occasionally employed in scenting soaps. Infusion of fruit is used as an enema for infants for the expulsion of flatus.

Fennel are valuable for relieving gas and expelling wind from the stomach. An infusion is highly beneficial in treating biliousness, flatulence and indigetion. Seeds are an effectve remedy for colic. The leaves and the seeds have mucus clearing property and hence promote the removal of catarrhal matter and phlegm from the brochial tubes. They are useful in bronchitis and asthma. An infusion of seeds is remedial in painful menstruation. Fennel tea is recommended for weakened, sore or inflamed eyes.

23. Fenugreek
Trigonella foenum-graecum Linn.
Fabaceae (Papilionaceae)

Fenugreek has long history as a medicine and food. Fenugreek seed is popularly known as Methe and its leaves (fresh or dried) as

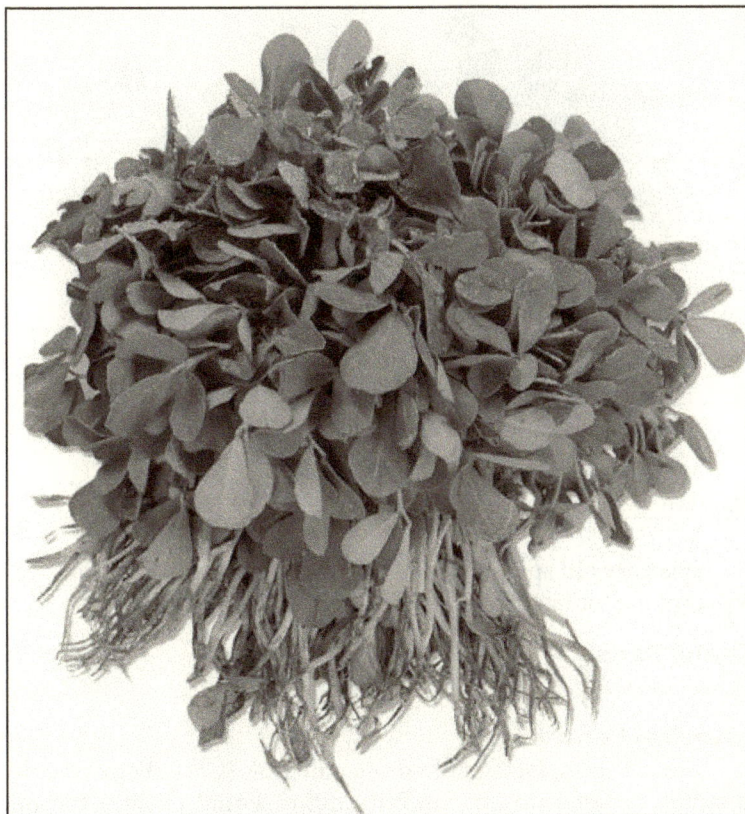

Methee or Methi. It is native to south-eastern Europe and West Asia, and now cultivated in India, Argentina, Egypt and Mediterranean countries (southern France, Morocco and Lebanon) also. The seed is small and yellowish-brown. It has a pleasantly bitter taste and a peculiar odour and flavour of its own. Medical Papyri from ancient Egyptian tombs reports that fenugreek was used in those days to reduce fevers and as a food. In religious rites, fenugreek was one among the components of Holy Smoke, an Egyptian incense, uses in fumigation and embalming. Roman history (second century BC) states that Porcius Cato (then Roman Authority in Animal Husbandry) ordered that fenugreek should be sown as a fodder for oxen in the Roman fields. In the Middle Ages, fenugreek was recommended as a cure for baldness in men.

SpiceFlair

It is very old spice of India. It is understood that this spice was a part of the Indian diet even 3,000 years ago.

Useful Parts

Leaves, Seeds.

Chemical Constituents

Seeds: Protin, fat, carbohydrate, minerals (Ca, P, Fe), carotene, thiamine, riboflavin, niacin, fatty oil, essential oil, volatile oil, trionelline, methybetaine of nicotinic acid, choline, saponin, etc.

Processed Products

1. *Fixed Oil*: The fixed oil content is about 7 per cent. The fatty acids consist largely of oleic and linolenic acids. It has marked drying properties. The dried oil is golden yellow in colour and is insoulble in ether. The oil has a disagreeable odour and bitter taste.

2. *Volatile Oil*: The volatile oil content of fenugreeek is very small. It is brown in colour and slightly odourous. Very little is known of its chemical composition. It is not of any commercial importance in India yet.

3. *Oleoresin*: Oleoresin of fenugreek is now exported. It is gaining importance.

4. *Methi Leaves*: Leaves contain vitamin C. After boiling in water or steaming and then frying, the vegetable loses of the vitamin. Results of organoleptic evaluation showed that the most acceptable method of preparation is by steaming.

Properties and Uses

It is used both as a food additive and medicine as well. Fresh tender pods, leaves and shoots which are rich in iron, calcium, protein, vitamin A and vitamin C are eaten as curried vegetable since ancient times in India, Egypt, etc. As a spice, fenugreek also adds to be nutritive value and flavour to foods. In the Middle and Far East it is a popular leaf vegetable where meatless diets are customary for cultural and religious reasons. Dried leaf powder is also used for garnishing and flavouring a variety of foods in India.

It is a popular ingredient of bread in Egypt and Ethiopia. It is known to the Arabs as 'hulba' and in Ethiopia going by the 'Abish'. In Greece, the seeds, boiled or raw, are eaten with honey. In the United States, seed is used in the manufacture of chutneys and in various spice blends. Its most important culinary use is as a source of fenugreek extract, the principal flavouring ingredient of imitation maple syrup. It is used in recipes like 'Hearty Vegetable' 'Bean Soup' and Fenugreek Beef Strew', etc. Fenugreek seed is one of the principal odorous constituents of curry powder. It is specially used in mango pickle in North and North West India, in particular. Fixed oil is used in perfume.

In Medicine

The seeds are used in colic, flatulence, dysentery, diarrhoea, dyspepsia with loss of appetite, chronic cough, drops, enlargement of liver and spleen, rickets, gout and diabetes. The seeds are used as carminative, tonic and aphrodisiac. Its infusion is administered to small-pox patients as a cooling drink. It is also used in sweets served to ladies during post-natal period.

Fenugreek seeds contain substantially the steroidal substance diosgenin which is used as a starting material in the synthesis of sex hormones and oral contraceptive. Nowadays, medicines in tablet form are manufactured in the UK, Mexico and Central America so as to make the birth control a success. As a medicine, fenugreek is also used for curing mouth ulcer, chapped lips and stomach ailments and for getting relief from gas troubles. It is beneficial in diabetes.

Ground fine and mixed with cotton seed, it is fed to cows to increase the flow of milk. Mildewed or sour hay is made palatable to cattle when fenugreek herbage is mixed with it. It is used as a conditioning powder to produce a glossy coat on horses.

In java, it is used in hair tonic preparations and as a cosmetic. The powder made from the seeds is used as a yellow dye; in the Far East. Harem women in North Africa and the Middle East use roasted fenugreek seeds to achieve a captivating buxom plumpness. Fenugreek oil gets an important place in growth of hair. Hair-tonics are manufactured from fenugreek oil. A paste of fenugreek seed is applied to the hair by all women before bath, once in a week, in South India. Eating fenugreek leaves fried with ghee is reported to give relief from giddiness, stomach pain, etc. Swelling on any part of the body can be reduced by using the paste of fenugreek leaves.

Seeds are useful for digestive systems disorders *e.g.* colic, flatulence, dyspepsia and dysentery. They are valuable in treating anaemia and fever. The respiratory tract ailments like bronchitis, influenza, sinusitis and catarrh are benefited by fenugreek tea. A gargle of seeds is the best for sore throat.

24. Galangal
Alpinia galanga (Linn.) Willd.

Zingiberaceae

The earliest reference to Galangal is that of an Arabian geographer Ibn Khurdabah (869-885 A.D.) He included galangal in a list of products from a country he refers to as Sila (China). Sometime later, Plutarch mentioned the use of galangal as a fumigating spice by the early Egyptians. In 1153, it was mentioned by Edrisi as an incoming spice into Aden, which was then the port used by the Egyptians for all Asiatic products. Marco Polo also mentioned that it was produced fairly extensively at Kachanfu and at Kinsai (Tonkin).

Galangal is one of the spices which reached the European markets relatively early; It was mentioned along with pepper in the literature of the Middle Ages in Europe where it received praise in various writings dealing with drugs, medicines, etc.

The source of the spice which was reaching the European markets shrouded with obscurity. In 1867, a plant of unknown

botanical origin was discovered in the extreme South of China. The following ear, Hance examined both dried and living specimens of the plant. He was able to conclude that it was the same as the 'lesser galangal' known to the pharmacists.

Galangal is the dried rhizome or root of a plant (A. *galanga*). It grows mainly in the eastern Himalayas and south-west India. The plant is 1.8-2.1 m. high, and bears perennial rhizomes. Rhizomes are deep orange-brown in colour, aromatic, pungent and bitter. The fruits are about 13 mm long, constricted in the middle, and contain 3-6 seeds. The latter are slightly pungent, with an aroma similar to that of the rhizome. Cut pieces of the rhizomes are called '*greater galangal*. The rhizome of 'lesser galangal' is smaller and reddish brown in colour, and has a stronger odour and taste. According to some, galangal root is also known under the names of 'smaller galangal' and 'Radix galangal minoris'.

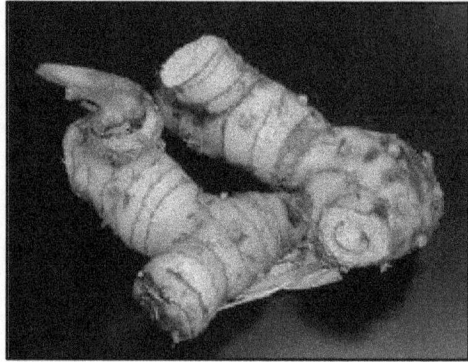

The botanical identity of 'Galangal' appeared shrouded in ancient past. This is indicated by its different names such as 'Greater Galangal', 'Smaller Galangal', 'Lesser Galangal', Chinese Galangal'. Interestingly, Galangal is referred to species of two different genera *viz.*, Alpinia and Kaempferia of the family Zingiberaceae. (The two genera are botanically clearly distinct). They are named as '*Alpinia galangal* (Linn.) Willd and *Kaempferia galanga* Linn. The present authors (Dr. Patil and Dr. Dhale) are hence inclined to state that the commercial samples of the drug still need correct botanical identity.

Useful Parts

Rhizomes.

Chemical Constituents

Green rhizomes, essential oil consisting methyl-cinnamate cineol camphor, and probably d-pinene, oleoresin.

Processed Products

Volatile Oil

It is steam-distilled from the dried comminuted rhizomes of galangal. It is a pale-yellow to olive-brown liquid with an eucalyptus cardamom-ginger-like odour and warming camphoraceous-like bitter taste.

In Medicine

In Malaysia, they are used as spice, and the fruits as substitutes for cardamoms. The drug has an important action on the bronchioles. The rhizome and its essential oil are useful in respiratory troubles, specially of children. The rhizomes are also carminative and stomachic.

Galangal Oil

It is used as a trace constituent in flavour studies. It is rarely used in perfumery. It could, however, be of value in an oriental or spice-type perfume formulation.

Galangal Oleoresin

It is used in medicine and in flavouring foods, etc.

25. Garlic
Allium sativum Linn.

Liliaceae

Garlic (lassan) belongs to the family Liliaceae. It has been recognized since time immemorable all over the world as a valuable condiment for foods and as a popular folk remedy.

It was consumed by Egyptian workmen 5,000 to 6,000 years ago. The Greeks loved it, while the Romans of Caesar's day, did not. Garlic is a native of Western Asia and Mediterranean area. It has been used in cultivation for centuries. Europeans, especially the Italians and Spanish, have used this regularly for 2,000 years and more. The Spanish are believed to have brought garlic to the New World. It became there an immediate favourite with Red Indians. In most Mediterranean countries and among the families of Italian and Spanish descents in the United States, garlic forms a part of almost every dish. Garlic is a hardy, bulbous, rooted perennial plant. It bears narrow flat leaves and small white flowers and bulbils. The compound bulb consists of 6-34 smaller cloves. The medicinal qualities of garlic have been celebrated since earlier times. Pliny has said that it is a remedy for 61 ailments, while Aristotle, Sotion and Dioscorides also have sung its praises in their writings. An English writer says that in India, garlic has been a favourite for centuries for "improving the voice, intellect and complexion, promoting the union of fractured bones, and helping to cure nearly all the ills which form from flesh to hair". Garlic possesses highly curative properties, it is

described as derived from Amrita or Ambrosia. The bias against the use of garlic in India, particularly among the Brahmins, appears to have originated from its popularity with the foreign invaders. The prejudice became so intense that not only the socio-religious writers like Manu deprecated its use, the authors of medicine like Kashyapa also disparaged it. The recent scientific evidence about its several highly curative properties clearly shows that the old bias was not justified. On the contrary it is one of the cheapest and most efficient folk medicines in the hands of housewife and the layman alike for day-to-day use, which should be encouraged.

No distinct varieties of garlic are recognized by the trade in North India, but two distinct types, namely Fawari and Rajalle Gaddi with slightly bigger bulbs are grown in the Bellar district of South India. The Indian varieties can be divided into two groups:

1. Small to medium-sized bulbs with small to large number of cloves, and
2. Big-sized bulbs and cloves.

Useful Parts
Bulblets.

Chemical Constituents
1. *Fresh Garlic*: Protein, fats, mineral matter, carbohydrates, calcium, phosphorus, iron, nicotinic acid, vitamin.
2. *Dehydrated Garlic*: Protein, fats, mineral matter, carbohydrates, calcium, phosphorus, iron, sodium, potassium, vit.-A, vit.-B, vit.-B$_2$ niacin, and vit.-C.

Processed Products
The various forms of dehydrated garlic are all produced by slicing peeled sound garlic cloves into flat slices, which are dehydrated, graded and further processed as necessary. The following broad categories are recognized in the trade.

1. *Dehydrated Garlic Slices or Rings*: Dehydrated garlic in pieces larger than 4 mm are obtained by cutting garlic cloves into slices and removing broken pieces, smaller than 4 mm by sieving.
2. *Garlic Flakes or Garlic in Pieces*: Dehydrated garlic passing through a sieve of 4 mm aperture size but retained on a sieve of 1.25 mm aperture size.
3. *Garlic Gritts*: Dehydrated garlic passing through a sieve of 1.25 mm aperture size but retained on sieve of 2.50 mm aperture size.

4. *Garlic Powder*: Dehydrated garlic passing through a sieve of 0.25 mm aperture size; a uniform product of which 95 per cent passes through the sieve.

5. *Garlic Salt*: Comprising 20 per cent garlic powder, 78 per cent table salt, and 2 per cent desiccant.

6. *Odourless Garlic Powder*: A patented process developed at the CFTRI, Mysore.

7. *Oil of Garlic*: Essential oil of garlic obtained by steam distillation.

8. *Garlic Oleoresin*: Garlic concentrate is manufactured by extraction of garlic with water by vacuum concentration, the expressed juice and the aqueous extract of the press-cake. Oleoresin garlic is described as a "dark brown soft extract". To confer homogeneity and ease of handling, the concentrate is usually compounded. Essentially, garlic extract contains sugar, water, minerals, nitrogeneous compounds and certain amount of aromatic materials. The essential oil constitutes the aroma; commercial oleoresin containing up to 5 per cent essential oil is available. There are several types of garlic oleoresin extractives depending upon the end-use.

Properties and Uses

Garlic has since long been reputed worldwide as a valuable condiment for foods, and also as a popular remedy for various ailments and physiological disorders. According to the Unani and Ayurvedic systems, as practiced in India, garlic is carminative and is a gastric stimulant and thus aids in digestion and absorption of food. It is also given in flatulence. In modern allopathy, it is being used in a number of patented medicines and other preparations. Besides, garlic is also anthelmintic and antiseptic. The active principle in garlic is an antibiotic allicin, which is an enzymatic cleavage product from its precursor, 'allin', naturally present in garlic. However, allicin is highly thermolabile (sensitive to heat) and should be protected from heat or high storage temperature. It is also very hygroscopic and hence great care is needed in its moisture-proof packaging.

Garlic as Condiment

In America, fresh garlic is dehydrated and used in mayonnaise products, salad dressings, tomato products and in several meat preparations. Further, raw garlic can be used in the manufacture of garlic powder, garlic salt, garlic paste, garlic vinegar, garlic cheese croutins, garlicked potato chips, garlic bread, garlicked meat tit-bits and garlicked bacon, etc. In India, and other Asian and Middle East countries, it is already being used in several food preparations, notably in chutneys, pickles, curry powders, curried vegetables, meat preparations, tomato ketchup and sauces. Garlic is used ground with other spices, sliced and fried or bruised and steeped in water, in vegetables, meat and fish preparations.

Garlic Oil as Insecticide

Garlic oil is considered to be an effective insecticide. The larvicidal principles of garlic have been isolated and identified as 'diallyl disulphide' and 'diallyl trisulphide' which are fatal. Garlic has been reported to be effective against many crop pests including certain beetles, mites, thrips, aphids, whiteflies, moths, nematodes, armyworms and borers.

Garlic Paste/Mixture as Biofungicide

A garlic remedy for the dreaded slow wilt disease in black-pepper has been described.

Garlic Oil as an Adhesive

Garlic oil is highly volatile. It is very useful as an adhesive for broken glass.

Garlic Residue with Antibacterial Properties

The residue of garlic, obtained by alcoholic extraction and distillation, contains a bacteriostatic and bactericidal substance, identified as 'allyl disulphide oxide'.

In Medicine

Garlic has been used as an attempted cure for drops, pherenitis, jaundice, pneumonia, scrofulous swelling of the neck, bronchitis asthma, tuberculosis and many other ailments. Externally, garlic is used as a remedy for skin and ear diseases. In western medicine also, it is used to a fairly large extent. Garlic juice mixed with 3 or 4 parts of ordinary or distilled water has been used as a lotion for washing wounds. The inhalation of garlic oil or garlic juice has been recommended in cases of pulmonary tuberculosis, rheumatism, sterility and impotency. Garlic juice is used for various ailments of stomach; as a rubefacient in skin diseases and as ear drops in ear-ache. The juice diluted with water can be used against duodenal ulcers. In Combodia, the leaves are used in the treatment of asthma.

In Ayurvedic system of medicine, garlic has been reported to be remedial in over different 18 diseases of human beings.

26. Ginger
Zingiber officinale Rosc.

Zingiberaceae

The Sanskrit name 'Singabera' gave rise to the Greek and to the Latin name *Zingiber*. The generic name Zingiber is also thought derived from Malayalam and Tamil names, 'Inchiver' and 'Inchi' respectively. The plant is thought originated in India and was introduced in China at a very early date. It has in vogue as a spice by

Indian and Chinese from early times. It is mentioned in many Chinese and Indian treatise. It was known in Europe in 1st century A.D. and was mentioned by Dioscorides and Pliny. It was brought by Aarab Traders from India. They also took it to East Africa in the 13th century and the Portuguese to West Africa and other part of the tropics in the 16th century. It was introduced in the New World very early and was exported from San Domingo (1585). Many cultivars of ginger grow in India. They are named after localites *e.g.* Maran, Ernad, Wynad, Kuruppampudi, Himachal, Nadia, etc. Rio-de-Janeiro is exotic cultivar.

Ginger is the dried underground rhizome of the plant. It constitutes one of the most important major spices of India. Ginger is one of the oldest spices of the world. The sundried rhizomes in trade are known as 'hands' or 'races'. They are either with outer brownish cortical layers intact or with outer coating partially or completely removed. It is a native of tropical South-East Asia. Ginger is cultivated in several parts of the world like India, Jamaica, Sierra Leone, Nigeria (West Africa), Southern China and Japan. It is also grown in Taiwan and Australia. Jamaica and India yield the best quality ginger, followed by West African variety. Chinese ginger is usually not exported as a dried spice, but preserved in sugar syrup or converted into 'ginger candy'. It is less pungent and aromatic. It cannot be therefore used for distillation or extraction purposes. Japanese ginger possesses a certain pungency, but is devoid of characteristic ginger aroma.

Useful Parts

Fresh and dried rhizome.

Chemical Constituents

Volatile oil, oleoresin, thiamine, carotene, Vit. A, B, B_2, C, niacin, minerals (Ca, P, Na, Fe, K).

Processed Products

Different products are manufactured from fresh and dried gingers. They are as follows:

(A) Dried Ginger Products

1. *Ginger Oil*: Dried and cracked and comminuted ginger yields 0.3 to 3.5 per cent of pale-yellow viscid volatile oil. The oil possesses aromatic odour but not the pungent flavour of spice. The odour of oil is quite lasting. Oil is generally extracted from unscraped dried ginger or from ginger scrapings. The Jamaican ginger yields about 1 per cent of oil, while African ginger and Indian ginger. The essential oil derived from dried ginger, known in trade as 'oil of Ginger' is greenish to yellow in colour, mobile with the characteristic warm and aromatic odour The aroma somewhat varies with the type of ginger used in distillation. The oil becomes viscous on storage and stored oil contains a high proportion of non-volatile constituents. The major constituent of volatile fraction in stored oil is curcumene, and in freshly distilled oil zingiberene. The oil distilled from dried rhizomes, which have been stored for long periods, has a higher density and is less laevo-rotatory than oil distilled from freshly dried rhizomes.

2. *Ginger Oleoresin*: Oleoresin is obtained by extraction of powdered dried-ginger using solvents like alcohol, acetone or any other efficient solvent like ethylene dichloride. It contains volatile oil and non-volatile pungent principles for which ginger is so highly esteemed. The yield and

quality of the oleoresin depends upon the raw material, solvent used and the method of extracting ginger oleoresin. It is commercially known as 'Gingerin'. It generally contains gingerol, zingerone, shogaol, volatile oil, resins, phenols, etc. Ginger oleoresin is manufactured on a commercial scale in India and abroad. It has great demand in food industries abroad and to some extent also in India.

The oleoresin contains essential oil and also non-pungent substances. The amount of essential oil in oleoresins in different ginger-growing areas varies from 7 to 28 per cent ; non-pungent substances may amount to 30 per cent. The amount of essential oil is an important factor in the evaluation of oleoresins. The non-pungent substances associated with oleoresin include carbohydrates, palmitic acid and other fatty acids. The pungent principles of ginger are oxymethyl phenols. Commercially, preparations of oleoresin contain gingerol,shogaol, zingerone and small quantities of paradol. The proportion of different constituents in oleoresin vary. Freshly prepared oleoresin contains gingerol as the main constituent (approximately 1/3 of the weight of oleoresin), whereas commercial samples which have been stored for long periods contain mainly shogaol. Shogaol and zingeone do not naturally occur in fresh rhizomes. The presence in the commercial preparations is due to chemical changes brought about during the preparation and storage of the oleoresin.

3. *Dehydrated Ginger*: The ginger is either (i) sun-dried or (ii) dehydrated mechanically. (i) Sun-dried ginger is available in two forms: coated or unscraped and uncoated or scraped. Green ginger is placed in a wire-gauge cage and dipped in a boiling solution of 20, 25 or 50 per cent sodium hydroxide for 5, 1 or 0.5 minutes. The cage is pulled out of the lye bath and rhizomes are placed in 4 per cent citric acid for two hours. They are washed and dried. Mechanical peeling of rhizomes is done using Hobart abrasive peeler. Abrasive peeling for 60 seconds gives a product equal in essential oil content to hand-peeled ginger. However, hand-peeling was found to be superior to mechanical peeling in giving a product.

4. *Bleached Ginger*: Green ginger is placed in large, shallow cisterns. Water is added until water level is above rhizomes by about 30 cm. The entire mass is trampled in order to clean rhizomes of soil and roots and to peel off part of the skin. This process is repeated. The peeled rhizomes are then repeatedly immersed in milk of lime and allowed to dry in the sun until the ginger receives a uniform coating of lime. It assumes a bright colour. The final drying takes about 10 days. Finally, the product is well rubbed with gunny cloth to remove remnants of skin and to provide a smooth finish. This is known as bleached or 'limed ginger'.

5. *Other Food Products*: The possibility of preparation of food products from whole ginger and ginger residue or spent ginger have suggested *e.g.* (i) starch from spent ginger (ii) vitaminized ginger powder cocktail (iii) vitaminized effervescent ginger powder for use as soft drink. These powders on dilution with chilled water yield delicious and refreshing drink. The vitaminized powder is sufficiently rich in vitamin C. The drinks prepared is refreshing, contains vitamin C sufficient enough to meet daily human requirements.

(B) Products from Fresh Ginger

Various products are prepared using fresh ginger rhizome *e.g.* (a) Ginger preserve of 'Murabba'. (b) Ginger candy. (c) Soft drinks like ginger cocktail which also aids in digestion. (d) Ginger pickles, salted, in vinegar or in vinegar mixed with other materials like lime, green chillies, etc. (e) Alcoholic beverages like ginger wines, ginger brandy and ginger beer.

Preserved ginger consists of peeled or unpeeled green rhizomes impregnated with sugar syrup. It can be converted into crystallized ginger by further processing. Fresh ginger can also be pickled in salt and vinegar or lemon juice as a preserve.

Properties and Uses

The pleasant and spicy aroma, the penetrating flavour, pungent and slightly biting due to antiseptic and pungent compounds present in it render it indispensable in the manufacture of food products *e.g.* ginger bread, confectionery, ginger ale, curry powders, certain curried meats, table sauces and in pickling. It is used in the manufacture of certain soft drinks *e.g.* cordials, ginger cocktail, carbonated drinks, bitters, etc. Ginger preserve is used chiefly in confectionery. Chocolate manufactures utilize the preserve for enrobing. It is also used in jams and marmalades. The syrup, in which ginger is preserved, is valued by pickle and sauce manufactures. It is also useful in making ginger bread. The production of preserve or ginger 'murabba' is a traditional industry in India. A number of alcoholic beverages are prepared from ginger such as ginger brandy, ginger wine, ginger beer and ginger ale. It is little used in perfumery.

In Medicine

Ginger is considered to be carminative and stimulant in Ayurvedic medicine. It is advised in dyspepsia and flatulent colic. It is also prescribed as an adjunct to many tonic and stimulating remedies. It also has aphrodisiac property besides its use in tinctures and as a flavourant. It is used in asthma, obesity, gingivitis, indigestion, etc. Tea made from pure roasted ginger offers a fragrant and refreshing beverage with useful medicinal properties. It induces relaxation and is a carminative to reduce flatulence. It is beneficial in clearing chest and nasal congestion in cold and influenza. The brew has been used for centuries by the Chinese especially after a heavy meal. Veterinary uses of ginger are carminative in indigestion of horses and cattle, in spasmodic colic of horses, and to prevent griping by purgatives.

27. Horse-Radish
Cochlearia armoracia Linn.

Brassicaceae (Cruciferae)

Horse-radish is one of the oldest condiments. Cultivation in the native area has been practiced for at least 2000 years and the plant was referred by Dioscorides. It is a well-known, large-leaved, hardy perennial in many long-established gardens. It is the thick, white, fleshy tasty root of the horse-radish which is highly prized as an appetizing condiment with certain foods.

It is a native of the marshy districts of Eastern Europe. It is grown in the USA, and to some extent in the gardens, both in the north India and in the hill stations of the south India. The tap-root is tuberous and cylindrical, 30-cm long and about 18 mm across. It

possesses an acrid, pungent taste and, when scraped or bruised, emits a characteristic pungent odour.

Volatile Oil

Distillation of the titrated roots gives about 0.05 to 0.2 per cent volatile oil, which is not produced on a commercial scale.

Useful Parts

Roots.

Chemical Constituents

Protein, fat, carbohydrates, and vitamin C, ascorbic acid, volatile oil (singrin) sulphur containng glycoside.

Propertis and Uses

It is used as appetizer condiment. It is digestive and anti-scorbutic because of its high vitamin C content. It is highly prized as a condiment, specially with oysters and meats. Leaves and roots are also used as food in German. A freshly grated horse-radish root and mixed with vinegar and salt, is much appreciated as an appetizing and pungent condiment. It enhances the flavour of the boiled or roasted beef. When mixed with ketchup, the grated root imparts a refreshing taste to seafoods, especially shrimp cocktail and oysters. Horse-radish is used for 'horse-radish cream' 'sauce' or 'relish'. It is rather a salad cream and contains at least 25 per cent shredded horse-radish (fresh) or equivalent dehydrated root, essential oil, cream containing at least 25 per cent edible vegetable oil, 7.5 per cent of nonfat milk solids (MFMS).

In Medicine

Horse-radish is stimulant, diaphoretic, diuretic and digestive. It is also used as a counterirritant in lumbago and similar painful afflications.

28. Hyssop
Hyssopus officinalis Linn.

Lamiaceae (Labiatae)

It is native to southern Europe and temperate zones of Asia. It has been naturalizd in the USA. Leaves are sessile, linear-oblong or lanceolate, obtuse, entire; flowers are bluish purple, in axillary tufts arranged unilaterally on terminal branches. Leaves and flower tops constitute condiment or spice. It is cultivated in Europe, particularly in southern France, mainly for its essential oil. It is also grown in gardens for ornamental purpose. It is a perennial shrub, 30-60 cm high, found in the Himalayas, from Kashmir to Kumaon at altitudes of 2,435-3,335 m.

Useful Parts

Leaves, Flowering tops.

Chemical Constituents

Herb: volatile oil, fat, sugar, choline, tannins, carotene, iodine, xanthophylls. The tops: ursolic acid, a glucoside (diosmin) which on hydrolysis yields rhamnose, glucose and a glucone, diosmetin (4-methyl luteolin). Hyssop is stated to yield a greyish green dye.

Processed Products

1. *Essential Oil*: The oil yields vary greatly with the condition of the plant material, depending chiefly upon its dryness. It averages from 0.15 to 0.3 per cent for fresh to clover-dry, and from 0.3 to 0.8 per cent for clover-dry to completely dried material. Maximum yield is obtained from plants harvested just after the opening of the blossoms. The oil obtained from all the above ground parts is of uniform quality. Hyssop oil is colourless or greenish yellow with an agreeable aromatic; somewhat camphoraceous odour, and warm, slightly bitterish taste. About 50 per cent of hyssop oil consists of ketone 1-pinocamphone.

Propertis and Uses

The leaves and flowering tops have an agreeable aromatic odour and a warm, pungent, bitterish taste. They are used as flavouring for salads and soups, and also in the preparation of liquors, bitters, tonics, and specially in liquors of French type, in Chartreuse and Benedictine, etc. and perfumes. The green parts are also used as pot-herb. Hyssop oil is used as a flavouring agent in bitters and tonics, especially in French liqueurs of the Chartreuse and Benedictine type. It is also used to some extent in perfumes with a spicy odour.

In Medicine

Hyssop is stimulant, carminative and pectoral, and used in colds, coughs, consumption and lungs complaints. An tea of infusion prepared from the plant is effective in nervous disorders and toothache and in pulmonary, digestive, uterine and unrinary troubles. Leaves are stimulating, stomachic, carminative and useful in hysteria and colic. The leaf juice is used for the expulsion of round worms. The crushed herb is applied as a resolvent and vulnerary. Infusion of plant is beneficial in asthma and coughs. Steeped in hot water, it is used as a resolvent and vulnerary, as fomentation for wounds, sprain and strains, muscular rheumatism and for clearing discolouration of skin due to blows. It is also administred as a salve in catarrhal opthalmia. The oil promotes expectoration in bronchial catarrh and asthma.

29. Cinnamon (Cassia)

Cassia (Jangli Dalchini) and True Cinnamon (Dalchini or Darchini) are very popular spices commonly used in the Indian cuisine. The name 'Dalchini' is a derivation from the Arabian term,

'Dar-l-chini', meaning the wood (or bark) of Chin, one of the oldest and largest producers of Cassia, so well known in commerce, is China. Cinnamon is one of the oldest spices known to mankind. It was well known to Egyptians even 2000 years before Christ. European and Arabian travellers and writers witness the trade in cinnamon and cassia from Ceylon, Seychelles, China, Vietnam, Madagascar, Indonesia and West Coast of India to other countries. Chinese medicinal treatise mentioned them in their earliest herbals as well. Even today cinnamon and cassia occupy an important position in the world trade. In pre-war years, trade in cassia was more important than that in cinnamon.

Cassia is generally cheaper and considered inferior to cinnamon. It is often used as its substitute. Cassia and cinnamon, when whole are very easily recognizable from each other. The former is with coarser skin of greater thickness and milder flavour. However, in the ground form, it is rather difficult to identify.

Cinnamon and cassia consist of layers of dried inner barks of branches of a number of species of evergreen tropical trees belonging to even the same species known by different names in different producing countries. However these species have been officially recognized as such: (a) Cassia China or Cassia *i.e. Cinnamomum aromaticum* C.G. Ness or *C. cassia*, Blume(b) Indonesian or Batavia Cassia or Java Cassia or Cassia vera or Padang Cinnamon *i.e. C.burmannii* Blume (c) Siagon Cassia (from Korea) *i.e. C. laureirii* Nees (d) Cinnamon or True Cinnamon *i.e. Cinnamomum zeylanicum* Blume.

The most important aromatic cinnamon and cassia barks which are in commercial use in India are: (i) the 'genuine' or 'true cinnamon' (*C.zeylanicum*); (ii) Karuva or 'Jangli-Darchini' (*C.inners*); Cassia species. (iii) Tejpat or 'tamal patra' or Indian Cassia lignea (*C. tamala*); Cassia species (iv) Tezpat or Ram tezpat (*C.obtusifolium*).

The international nomenclature, description and uses of the above four species are briefly described in the following:

29a. Cinnamon (True Cinnamon)
Cinnamomum zelanicum Blume

Syn. *C. verum*
Lauraceae

Since the early days of civilization Cinnamon was one of the first spices prized and enjoyed by man. It was thought precious as a

flavouring agent for food, as medicine, perfume and as one of the aromatics burned as incense. The Egyptians were importing cinnamon nearly 2,000 years before Christ. Romans were luxuriated in cinnamon scented baths. Ever medieval magician kept cinnamon on hand. It was one of the ingredients of 'love-potions'. The Chinese mentioned it in their herbals, and ever now it is in vogue medicinally.

Cinnamon is one of the most important tree spices of India. Like its cousin cassia, cinnamon consists of layers of dried pieces of the inner bark of branches and young shoots. It is obtained after removal of cork and cortical parenchyma from the 'whole bark'. The thickness of the bark varies from 0.2 to 1.0 mm. Pure cinnamon is superior to the cassia in appearance, flavour and odour. Cassia is the general substitute for cinnamon. One can make out these morphologically one from the other in the whole form, but it is difficult to do so in the powder form.

Useful Parts
All parts, Bark.

Chemical Constituents
Protein, fats, carbohydrates, minerals (calcium, phosphorus, iron, sodium, potassium), vit. B, vit B$_2$ vit. C, vit. A.

Processed Products
The commercial products of the cinnamon and cassia trees are in the form of: (i) whole and ground cinnamon/cassia, (ii) essential oils, and (iii) oleoresins.

1. *Cinnamon/Cassia Bark Oil*: It contains 0.5 to 1.0 per cent volatile oil. The essential oil is steam-distilled mainly from cinnamon chips and residue left after the preparation of quills for the spice trade. Bark oil is light yellow in colour when freshly distilled. It changes to red on storage. It contains cinnamaldehyde (60-75 per cent), eugenol and benzaldehyde, etc.

2. *Cinnamon-Leaf Oil*: The oil is obtained after steam distillation. It is heavier than water and can be easily separated from water by the decantation and filteration process. The oil is corrosive in nature and hence is packed and stored in dark-brown glass containers. It has a pungent odour, hot taste and contains eugenol (70-80 per cent) and traces of cinnamic aldehyde. The oil is yellowish with a slight camphoraceous odour resembling to that of clove oil.

3. *Cinnamon and Cassia Oleoresins*: Bark is distilled for the preparation of oleoresin. It is a dak brown liquid containing (50 per cent) volatile oil and has to be diluted before using

it as a flavouring agent. *Cassia oleoresin* is normally produced from Indonesian cassia which is much cheaper than Chinese cassia. Oleoresins are mostly manufactured in Western Europe and the USA.

4. *Other Products*: Some other products, almost minor, also found in the market are : (i) root bark oil, (ii) cinnamon seed oil, (iii) cinnamon buds for flavouring and spicing goods like quills and quillings.

Properties and Uses

All parts of the Cinnamon tree, *viz.,* bark, wood, leaves, buds, flowers, fruits and roots are put in use for various purposes worldwide and hence it is rendered very useful spice tree.

1. *Stem Bark*: It is widely used as a spice or condiment in the form of small pieces or powder. It is aromatic, astringent, stimulant and carminative and also possesses the property of checking nausea and vomiting. Cinnamon is used for flavouring confectionery, liquors, pharmaceuticals, soaps and dental preparations. Powdered cinnamon is a constituent of chocolate preparations made in Spain. Cinnamon is also used in candy, gum, incense, soaps and perfumes. Cinnamon has also been found to help diabetics in digestion of sugar.

2. *Stem Bark Oil*: It is extensively used for flavouring confectionery, liquors, pharmaceuticals, soaps and dental preparations. It has the cordial and carminative properties of cinnamon without its astringency and is employed as adjuvant in stomachic and carminative medicines. As a powerful local stimulant, it is sometimes prescribed in gastrodynia, flatulent colic and gastric debility.

3. *Cinnamon Leaf Oil*: It is useful in perfumery, cosmetics and flavouring industries. In the USA, the cheaper Seychelles oil is used as a source of eugenol in the synthesis of vanillin, while Ceylon oil, which is considered superior, is employed for perfuming soaps. It is also used to flavour sweets and confectionery. It is used as an embrocation in rheumatism.

4. *Seed Oil*: Seed contain fixed oil. It was formerly used for making candles. The oil, also called 'cinnamon suet' is

obtained by boiling crushed ripe fruits suspended in water. The oleaginous matter rises to the surface and solidifies on cooling.

5. *Cinnamon Buds*: They are useful for flavouring and spicing as the bark itself.

29b. Jangli-Darchini (Cassia or Karuva)
Cinnamomum inners Reinw.

Lauraceae

Cassia cinnamon (*Cinnamomum aromaticum* Nees. [syn. *C. cassia* Bercht and Presl.]), the true cinnamon, is a closely related to species *viz.*, *C. verum* Bercht and Presl. The products of commerce are Cassia bark, bark oil, leaf oil and oleoresin and are important in the international trade. Cassia is also obtained from various sources *e.g.* *C. cassia* Bercht and Presl., *C. burmannii* (C.G. and Th. Nees) B1. *C. laureirii* Nees and *C. tamala* Nees. These dried inner barks have a more intense aroma with high essential oil content (1 to 4.5 per cent as against 0.5 to 3 per cent in cinnamon bark). Unlike cinnamon, different parts of *C. cassia* contain only one type of essential oil in which cinnamaldehyde is the main component. Leaf-oil content of

cassia ranges from 0.7 to 2 per cent. Cassia is cultivated mainly in China.

The end use of the bark is the same as that of cinnamon reported later. The bark contains 0.5 per cent volatile oil with the odour of cloves. Its wood is light brown, fairly hard and shining. It machines well, takes on a good finish and hence can be used for the manufacture of light furniture.

29c. Tamal Patra or Tejpat
(Indian Cassia Lignea)
Cinnamomum tamala Nees and Eberm

Lauraceae

Tejpat attains a height up to 8 m and a girth of 1.4 m. It is a moderate-sized evergreen tree. It is distributed in tropical Himalayas, Khasi and Jaintia Hills and in eastern Bengal, Meghalaya, Assam, Manipur, Arunahal Pradesh, Tamil Nadu and Karnataka. The tejpat leaves are used widely in India as a flavourant or spice. The bark of the tree, known in the trade as Indian cassia bark or Indian cassia Lignea, is collected from trees growing at the foot of the Sikkim Himalayas. There are regular plantations of *C.tamala* in Khasi and Jaintia Hills, Garo Hills, Mikit Hills, Manipur and Arunachal Pradesh. Tejpat is grown mainly in the Jaintia Parganas of Sylhet

district. Many plantations in this tree are self-sown, a few are planted.

Useful Parts

All parts, Bark.

Chemical Constituents

Linalool, eugenol and cinnamic aldehyde, a-pinene, b-pinene, limonene and camphene.

Bark Oil

The essential oil from the bark is pale yellow, and contains 70–75 per cent cinnamic aldehyde.

Properties and Uses

The leaves are used chiefly as a spice or flavourant in various culinary preparations. In Kashmir, they are used as a substitute for betel (pan) leaves. It takes in India cookery the place of bay leaves, used in Europe. It is used as a clarifier in dyeing with myrobalans (kalala). The bark is aromatic. It is coarser than the bark of true cinnamon (*C. zeylanica*), and is one of the common adulerants of true cinnamon.

In Medicine

The leaves of C. *tamala* are carminative and are administered in colic and diarrhoea. They are reported to be hypoglcaemic, stimulant, carminative, antirheumatic, anti-diarrhoeal and are antidote for scorpion-sting.

29d. Tezpat
Cinnamomum obtusifolium Nees

Lauracee

It is a large robust tree, occurring in the central and outer eastern Himalayas, Bangladesh, Assam and Andaman Islands.

Useful Parts

Leaves, Root bark.

Properties and Uses

The leaves, also known as tejpat, are aromatic and have similar uses as those of *C.tamala*. The root bark is especially collected in Marataban. It resembles Ceylon cinnamon.

In Medicine

It is used in dyspepsia and liver complaints in Nepal.

Commercial Types and Classification

This is classified basically in two types:

1. *China Cassia (Cassia Lignea)*: This is the bark of the branches of *Cinnamomum cassia* Blume. It consists of bark in tubular form or in simple, single quills or in compound, double quils, as in the case of type Sri Lanka (Ceylone) cinnamon.

2. *Type Vietnam Cassia*: It occurs as greyish-brown bark in single or double quills. It varies in length (150 to 300 mm) in diameter (10 to 38 mm) and thickness (6 mm). Type Vietnam Cassia is classified in four grades such as thin, medium and thick rolls, besides broken pieces.

30. Japanese Mint
Mentha arvensis Linn.

Lamiaceae (Labiatae)

Mints are a group of plants which yield essential oil on distillation. Pudina or Japanese mint (*M. arvensis*) is well known to all in India, as used in 'chutney' and as an old popular household remedy for relieving cold and cough. In India, about 08 species of *Mentha* are reported to occur. However, the world demand for peppermint oil and menthol is met from the three species:

1. Mint or Japanese mint (*Mentha arvensis* Linn.) subsp. haplocalyx Briquet.

2. *M. piperita* var. piperita

3. *M. spicata* Linn. (syn. *Mentha viridis* Linn.)

The Greeks and Romans were quite familiar with *Mentha arvensis* (Japanese mint). Since ancient times, it was believed that Mentha (the damsel lover of God Pluto), was transferred into this plant due to anger of Prosarpain (the wife of Pluto and Goddess of Wealth). It is, therefore, this plant is commonly known as Mentha (in Latin). The Greek physicians, Mohammaden physicians and even Chinese and Japanese knew it since 2000 years ago. Japanese mint is a perennial herb. It has creeping rootstock and erect stem, 1-2 quadrangulate and branched, The plant rises to a height of 0.4 to 0.8 m, branching freely, beginning from its base. Flowers appear in May-June and again in September to November under cultivation. Leaves are 2.5-5 cm in length, oblong-ovate, or lanceolate. The various types of mints, which are cultivated commercially in India are:

1. Japanese mint (*Mentha arvensis*)

2. Peppermint (*M. piperita*)

3. Bergamot mint (*M.citrata*)

4. Spearmint (*M. spicata*)

Useful Parts

Entire herb, Leaves.

Chemical Constituents

Protein, fats, minerals (Ca, P, Fe), carotene, thiamine, riboflavin, niacin, vitamins - C, D and E.

Processed Products

1. *Essential Oil*: The younger and tender organs contain the largest amount of oil. With the advancement in the age of the leaf and increase in its size, oil does not increase proportionately because the number of glandular hairs remain the same, except that they are filled fully in mature leaves. Distillation of dried leaves is cheaper than that of fresh leaves. By steam distillation and filtration, a golden yellow volatile oil is obtained. Leaves and flowering-tops give the highest yield. About 50 per cent of menthol is separated out in crystalline form on cooling the oil. The remaining (dementholized) oil is used as peppermint oil.

   The natural oil yields 40-50 per cent of menthol and 50-60 per cent of dementholized oil. It can be used both in confectionery and medicine in place of imported peppermint oil. Japanese mint oil is not distinguished from the peppermint oil in the Indian trade. The dementholized oil has been found to contain menthol acetate (24.4 per cent), free menthol (44.6 per cent), menthone (24.6 per cent) and hydrocarbons (6.2 per cent). Among the hydrocarbons, alpha-pinene, alpha-1 limonene, carophyllene and cademens are present.

Propertis and Uses

Mint (Pudina) is very popular for flavouring chutney and vegetable. It is also used for flavouring meat, fish, sauces, soups, stews, vinegar, teas, tobacco and cordials. The fresh leaf-tops of all the mints are employed in beverages, fruit cups, sauce, ice-cream, jellies, salad, sauces for fish and meats. Roast lamb and mint jelly have become inseparable companions. Japanese mint oil is used as a substitute for true peppermint oil (*M.piperita*), which resembles it in physico-chemical properties. It possesses a somewhat bitter flavour and is considered inferior to *M. piperita* oil in aroma and quantity. Arvensis oil with low menthol content finding some use in cheap perfumery.

In Medicine

It is administered in a number of ailments *e.g.* for stomach disorders, in ointments for headaches, rheumatism and other pains, in cough drops, inhalations, mouth washes, toothpastes, etc. The oil and the dried plants are antiseptic, carminative, refrigerant, stimulant and diuretic. The dried plant is expectorant, emmenagogue, tonic to the kidney, useful in the disease of the liver and spleen, asthma, etc. It also possesses anti-spasmodic properties. It is used in jaundice, and frequently given to stop vomiting. In China, the leaves and stem are made into an infusion and used as carminative sudorific and anti-spasmodic. In Assam, the plant is considered as an excellent diaphoretic. An infusion is given in fever, indigestion, etc. Being stimulant, it relieves flatulence. It has anti-gas activity. It helps dissolve gravel in the kidneys and bladder. It is useful in pimples dryness of skin, dysmenorrhoea, pyorrhea, etc.

31. Japanese Star-Anise
Illicium religiosum Sieb *et* Zuc.

Magnoliaceae

The Japanese star-anise is known as 'Shikimi' in Japan. It is an evergreen tree with a trunk of about 3 m high with pale, yellowish white blossoms. The tree grows wild in warm localities of southern and central Japan, on the Loochoo Islands and in Formosa. It is grown extensively in the Prefecture of Nagasaki (chiefly on Goto

Island), and to a smaller extent in the Prefectures of Kochi and Tokushima on the island of Shikoku. The Japanese have been planting star-anise tree in temple compounds from a long time and also in cemeteries in order to protect them from desecration by wild animals. The fruits of the tree are poisonous and leaves emanate a peculiar odour which is supposed to keep animals away. The custom has developed to such an extent that even today, altars during funeral services are decorated with the leaves.

Useful Parts

Fuits.

Processing Technology

The dried fruits contain about 1 per cent of volatile oil with an unpleasant odour, quite different from that of the Chinese product (*Illicium verum* Hook. *f.*). The oil is not produced commercially yet. It is reported to be toxic. Its toxicity has been found to be due to the occurrence of a poisonous principle or a compound Henanomin. Its chemical composition is quite different from that of Chinese star-anise.

Propertis and Uses

In Japan, the fruit is gathered for use in incense sticks and for consumption as spice.

In Medicine

It has stomachic and tonic properties. Owing to its toxicity, it can be consumed in very small quantities only.

32. Juniper
Juniperus communis Linn.

Pinaceae

The common Juniper is an evergreen shrub. It sometimes attains the height of a small tree up to three metres with erect trunk and spreading branches, covered with a shreddy bark. The fleshy berry-like fruit does not ripen until second year. The roundish fruit is sub-globose, bluish black, dark-purple when ripe, 10-13 mm in diameter, covered with a waxy blook. The three scales comprise the fruit and occasionally make gap and expose the seeds. Seeds are bony usually 3, elongated, ovoid and embedded in pulp.

The plant is variable with a number of geographical varieties and garden forms. It becomes prostrate, not more than 60-90 cm in height at higher altitudes in the Himalayas. It runs wild in many parts of Europe and Asia, ranging as far east as the western

Himalayan mountains and north-eastern Asia. In North America, the bush or tree occurs south to Pennsylvania, and west to Illinois, New Mexico and northern California.

Useful Parts

Berries/Fruits.

Chemical Constituents

Fruits: Volatile oil, fermentable sugars, resin, juniperin probably a mixture of tannin and sugars, fixed oil, protein, wax, gum, pectins, organic acids (formic, acetic, malic, oxalic and glycolic), potassium salt.

Processed Products

1. *Juniper Berry Essential Oil (Volatile Oil)*: It is obtained by steam distillation of ripe fruits. The berries contain from 0.5 to 2.0 per cent essential oil. Ripe (blue) berries yield more essential oil than unripe (green) ones. The quality and yield of oil depends greatly upon the geographical origin of the berries. The oil from green unripe fruits is inferior; and in over-ripe fruits, the oil changes into resin. The bulk of commercial oil is obtained as a by-product during distillation of alcoholic beverages. The oil, however, is partly deprived of natural oxygenated odoriferous compounds.

   Juniper oil is colourless or pale-greenish-yellow limpid liquid with a characteristic odour of the fruit and with somewhat burning bitter flavour. On storing, the oil turns viscous and acquires a turpentine odour.

   The oil contains a-pinene which is the major constituent, others are found in smaller quantity *e.g.* camphene, cadinene, juniper camphor (possibly a sesquiterpene alcohol), a hydrocarbon (Junene) with strong diuretic properties, terpinenol, certain unidentified oxygenated compounds possessing characteristic juniper odour, and traces of esters.

2. *Manufacturing Alcoholic Beverages*: European countries like Hungary, Czechoslovakia and Ugoslavia, large quantities of Juniper berries are used domestically for fermentation and subsequent distillation of popular alcoholic beverages like 'Borovicka' It is a 'gin' type. In Germany. Austria and

Switzerland, the same drink is called 'Steinhages'. This spirit is nothing but an alcoholic distillate of fermented juniper berries containing as much as 40-50 per cent or more of alcohol. Slovaks are particularly fond of this strongly flavoured beverage.

Proprties and Uses

Fruits are sometimes used as an article of food. However, large quantities of fruits are used in Europe for preparation of alcoholic beverages of gin type It is largely used in compounded gin flavours, liquors and cordials. A twice rectified oil has high flavour value. Imitation-juniper oils have been produced.

In Medicine

The fruits and volatile oil are carminative, stimulant and diuretic. They are useful in different forms of dropsies, especially in conjunction with other drugs. They have been used in various oilments of urinogenital tract *e.g.* gonorrhoea, gleet and leucorrhoea, and in certain cutaneous diseases. Volatile oil of Juniper berries has been known for a long time as diuretic.

Other Uses/By-Products
Exhausted Fruits/Berries

The exhausted fruits (left after the distillation of oil) on concentrating repeated extraction with warm water yield (30-38 per cent) a product known as 'Succus Juniperi'. The preparation, consisting chiefly of invert sugar, was formerly used in Europe as a diuretic.

33. Kokam (Kokum) *Garcinia indica* Choisy

Guttiferae

Kokam (Kokum) is also known as 'Kokam Butter Tree', 'Brindonia Tallow Tree' or 'Mangosteen Oil Tree'. It should not be mistaken for mangosteen; (*Garcinia mangostana*). Kokam is a slender evergreen tree. Fruits are globose or spherical, 2.5-3.75 cm in diameter, dark purple when ripe, enclosing 5-8 large seeds. The tree is found in Southern India particular in Karnataka and the tropical rain forests of the Western Ghats, Konkan, Coorg and Wynad. It is planted in the southern districts of Maharashtra, notably in Ratnagiri and to

some extent in Gujarat. It flourishes well on the lower slopes of the Nilgiri Hils, West Bengal and Assam. It flowers from November to February and fruits ripen in April-May.

There are several trade varieties of Kokam available in Indian markets, *viz.* (i) Plain *Kokam*, (ii) Salted *Kokam*, (iii) Lonawala Kokam, (iv) Pakali Kokam, (v) Khane (edible) Kokam and (vi) Khoba Kokam.

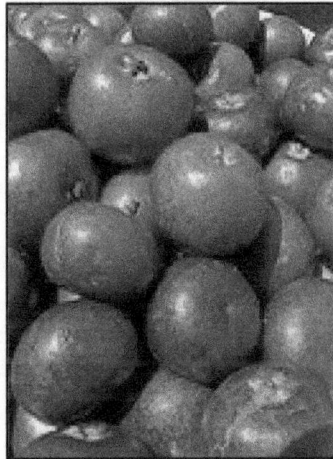

Useful Parts

Fruits.

Chemical Constituents

The composition of fresh Kokam rind comprising 50 - 55 per cent of fruit is as follows. Protein, tannins (polyphenols), pectin, starch, crude fat (hexane extract) acidity (hydroxyl-citric acid): pigment (anthocyanins), ascorbic acid, carbohydrates (excluding starch and pectin).

Kokam seeds (comprising 8-10 per cent of the whole fruit) contain about 25-30 per cent fat which is greasy and whitish yellow.

Processed Products

The fresh fruits are cut into halves and the fleshy portion containing seeds (8-10 per cent) is taken out. The rind constitutes about 50-55 per cent of the whole fruit. The rind contains about15 per cent acid (hydrox-citiric acid). The Kokam is prepared by sun-drying the rind of the ripe fruits after repeatedly soaking it in the juice of the pulp. The product so dried constitutes as the unsalted Kokam of commerce. Sometimes, after treatment with common salt during soaking and drying, the salted Kokam is produced. About 6-8 days are required for complete sun-drying.

Kokam seed is a good source of fat (about 25 per cent 30 per cent), which is known as 'Kokam Butter' in commerce and is described as follows:

1. *Kokam Butter:* The seeds yields fat. It is a valuable edible fat known in commerce as 'Kokam butter'. It is extracted mostly by crushing the kernels, boiling the pulp in water and skimming off the fat from the top; or by churning the crushed pulp with water. Oil is obtained by solvent extraction of seed-kernels in mechanical or motorized decorticators.

2. *Stearic Acid from Kokam Fat:* Kokam butter, like other Garcinia fats, is rich in combined stearic and oleic acids. It contains about 75 per cent of mono-oleo-disaturated glycerides. A method has been developed for the production of stearic acid from the fat with a yield of 45 per cent. It may be employed in the sizing of the cotton yarn.

3. *Kokam Syrup:* Kokam syrup is prepared from Kokam pulp/ juice mixing with sugar and water and used in medicine.

4. *Kokam Concentrate:* Kokam concentrate is used as an acidulant in various food preparations, curries, chutney, etc.

5. *Kokam Rind Acid:* The isolation of the hydroxyl-citric acid (15 per cent) in the pure form is carried out on a small scale. The pure acid finds a potential application in medicine.

6. *Kokam Colour*: Kokam rind contains 2-3 per cent anthocyanin pigments. It constitutes a promising source of natural colour for acidic foods.

7. *Ointment for Treating Carbuncles*: An ointment made out of Kokam fat, 'white dammar resin' (exuded by *Vateria indica* tree) and wax is said to be effective in treating carbuncles.

8. *Gamboge*: The commercial gamboge is a resin prepared from Kokam rind which was used extensively in the past in paints and varnishes.

Propertis and Uses

The fruit has an agreeable flavour and a sweetish acid taste. It is traditionally used in Konkan area of Maharashtra (India) chiefly in the form of Kokam. Kokam contains about 10 per cent malic acid and a little of tartaric or citric acid. It contains 15 per cent hdroxyl-citric acid. It is used as a garnish to give an acid flavour to curries and also for preparing cooling syrups during hot months. Italy and other foreign countries are also importing Kokam butter from India for use in confectionery. Kokam butter, as sold in the market, consists of egg-shaped lumps or cakes of light grey or yellowish colour with a grease feel and a bland oily taste. It is used mainly as an edible fat. It is also used as an adulterant of ghee. As ordinarily met with, it contains seed particles as impurities. Refined and deodorized fat is white and compares favorably with high-class hydrogenated fats. It is also suitable for candle and soap manufacture.

In Medicine

The fruit is anthelmintic and cardiotonic and useful in piles, dysentery, tumours, pains and heart complaints. Syrup from the fruit juice is given in bilious affections. Kokam butter is considered nutritive, demulcent, astringent and emollient. It is suitable for ointments and suppositories. It is used as a local application to ulcerations and fissures of lips, hands, etc.

34. Large Cardamom (Nepal Cardamom)
Amomum subulatum Roxb.

Zingiberaceae

Large Cardamom (Nepal Cardamom or Greater Indian Cardamom) is cultivated in swampy places along the sides of mountain streams in Nepal, Bengal, Sikkim and Assam. It is one of

the important cash crops of the eastern India. The leaves are oblong lanceolate and glabrous on both the surfaces. The plants are usually grown along small springs, in moist and shady sides of mountain streams and along the hilly slopes, usually at an elevation of 765 to 1,675 above the mean sea level. The plants mature during the thrid year of their growth, when flowers and fruits are produced.

Prevalent Varieties

Three well-recognized varieties of large cardamom in Sikkim are 'Ramshai', 'Sawaney' and 'Golshai' or 'Golse'. Some other known varieties, *viz.*, 'Ramla' 'Chibe', 'Ramnag', 'Madhusai' and 'Mongney' are derivatives of aforesaid varieties. Same variety is called by different names in different localities as variation in growing conditions may cause change in plant colour, whereas harvesting at different stages of fruit development may change fruit-skin colour. The major varieties grown in Sikkim are: **(i)** Ramshai or Ramsey: It is a Bhutia word which means yellow colour (Ram-colour, shai or sey-yellow). Plants are tall (approx. 2.5 m.) shoots are thin with long leaves. The fruits are smallest and are of inferior quality. The variety is mostly grown on higher altitudes aroudn 1,500 m. and above. (ii) Sawaney: It is a Nepali word which means that the variety is harvested in Sawan, *i.e.* August. Plants are also tall (approx. 2.5 m). The leaves are wider and shorter than to that or Ramshai. The fruits are bold and brown. This variety is popular in lower altitudes (below 1, 5000 m.) (iii) Golshai or Golse: It is a derivative of the words *'gol' meanings* round (Hindi) and 'Shai' *Isey*, yellow (Bhutia), indicating round shape and yellow colour fruits. The plants are short with 5-6 shoots. Leaves are also comparatively very short and broad. Fruits are big and bold.

Useful Parts

Capsules and seeds.

Chemical Constituents

Essential Oil (cineole, terpinene, terpeniol, sabinine, terpinyl acetate, bisabolene), polymerised oil.

Processed Products

Greater cardamom has only 3-3.5 per cent of volatile oil. It is not manufactured commercially. The method of drying greater cardamoms by smoking is time-consuming and unsatisfactory.

Properties and Uses

The ripened fruits considered to be a delicacy, are eaten raw by inhabitants of Sikkim and Darjeeling in September and October. Greater cardamom is used whole as flavourant in special vegetarian and non-vegetarian curried dishes like pulao, biryani and meat preparations. Its powder is used in curry powder and spice 'masala' mixtures. To some extent, it is also used in the manufacturer of essential oil and oleoresin.

In Medicine

The seeds are tonic to heart and liver, astringent to bowels, hypnotic, appetizer and cause belching. The outside covering is good for headache and for the teeth. It heals stomatitis. The decoction of seeds is used as a gargle in affections of teeth and gums. In combination with seeds of melons, they are used as a diuretic in case of kidney stones. They promote elimination of bile and are useful in congestion of liver. The seeds are used in gonorrhoea and as anphrodisiac. They are useful in neuralgia in large doses in conjunction with quinine. The seeds are an antidote to snake venom or scorpion venom. The seeds are also widely used in India as a spice or condiment, and in the preparation of sweetmeats. They are fragrant adjuncts to other stimulants, bitters and purgatives. An oil extracted from seeds is applied to eyelids to allay inflammation. It is aromatic, stimulant and stomachic.

35. Laurel Leaves (Bay Leaves)
Laurus nobilis Linn.

Lauraceae

It is also known by other names *viz.*, Sweet Bay. It is an evergreen shrub, small tree. It is grown in the Mediterranean countries and is cultivated in Greece, Spain, Portulgal, Asia Minor and Central America. It is sometimes grown in Indian gardens. The surface colour of the leaf is green; the underside is pale-green and somewhat yellowish. Laurel leaves are used whole or cracked. The aroma of the crushed leaves is delicate and fragrant. They are aromatic and bitter. Wreaths of laurel leaves were used by Greeks and Romans to honour their heroes. The laurel leaves originate from an evergreen hard tree or bush; cultivated since antiquity in Mediterranean countries. Size of the leaves ranges from 2.5 to 7.5 cm or more in length and 1.6 cm to 2.5 cm or more in breadth, at the widest part of

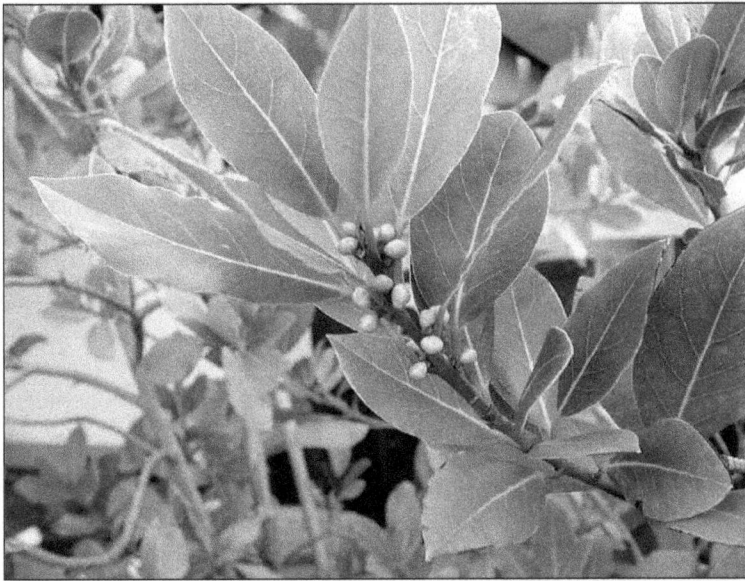

leaf. The leaf is elliptical, tapering to a point at the base. Dried berries, commonly called bay berries, have been imported into India for medicinal use. The berry is ovoid (about 1.5 cm long), black, coarse wrinkled and contains a single seed.

Useful Parts
Leaves.

Chemical Constituents
Dried bay leaves: protein, fat, carbohydrates minerals (Ca, P, Na, Fe) vit. B_1 (thiamine), vit. B_2 (riboflavin), niacin, vit. C (ascorbic acid): and vit. A, pentosans, cineole, fixed oil, volatile oil.

Processed Products
1. *Essential Oil*: The leaves on steam distillation yield essential oil with a characteristic sweet and spicy odour, reminiscent of cajuput. Fresh leaves and terminal branchlets yield 0.5 per cent oil, while dried leaves yield about 0.8 per cent. Principal constituent are cineol, which is a colourless liquid with a strong aromatic, camphoraceous odour and a cooling taste. Other organic compounds include a-

pinene, a-phelandrene, 1-linalool, 1-b-terpineol, geraiol, eugenol, eugenol acetate, methyl eugenol, a number of esters and acetic, isobutryric and isovaleric acids.

Propertis and Uses

Laurel leaves are used principally in vinegar pickle when packing pig's feet and lamb and pork tongue. They are also used in flavouring soups, stews, meat and game-dishes, fish and sauces, pickling spice, and in confectionery also. Laurel leaves are available whole or cracked, are not usually ground. Fat obtained from berries is used in perfumery. Berries are used in diarrhoea and drops. In Europe, they are also used to promote miscarriage.

In Medicine

Both leaves and fruits, possessing aromatic, stimulant and narcotic properties, were formerly employed for hysteria, amenorrhoea and flatulent colic. They are even used internally, though rarely, at present. Externally, however, commercial oil of laurel berry is sometimes applied as a stimulant in sprains, but its principal use is in veterinary medicine.

36. Leek (Winter Leek) and Stone Leek (Welsh Onion)

1. *Allium porrum* Linn.
2. *Allium fistulosum* Linn.

Liliaceae

Leek is non-bulbous member of the onion family and is grown for its blanched stems and leaves. Stone leek, Welsh onion or Ceboule or Japanese bunching onion, is *Allium fistulosum*. Winter leek is *A. porrum*. Sometimes Welsh onion is used as a substitute for leek. The white-stemmed stone leek is a native of eastern Asia, and was indroduced and domesticated in China and Japan. *A. fistulosum* has been mentioned in Japanese literature as early as 918 A.D. Today this bunching onion (Welsh onion) is cultivated in Siberia, China and Japan. More recently, it has been introduced into Europe and then to Russia in 1956. It is grown in home gardens in Europe, the USA and India. It is not grown on a commercial scale in India but is a favourite vegetable in kitchen-gardens.

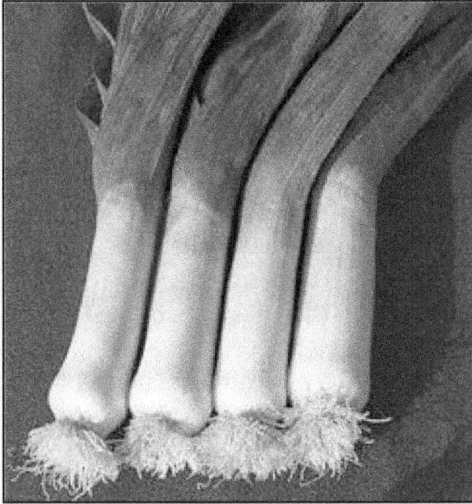

***Allium porrum* Linn.**

Stone leek (Welsh onion) is a hardy perennial. It is but grown as an annual or biennial. It does not form a real bulb but develops a small enlargement at the base. It may be propagated by division or seeds.

Useful Parts

1. *Winter leek*: Blanched stems and leaves.

2. *Stone leek*: Bulblike enlargement at base of stem and leaves.

Properties and Uses

Stone leeks are used for flavouring or for livening dull starchy diets of eastern peoples. They are also used in salads or eaten raw

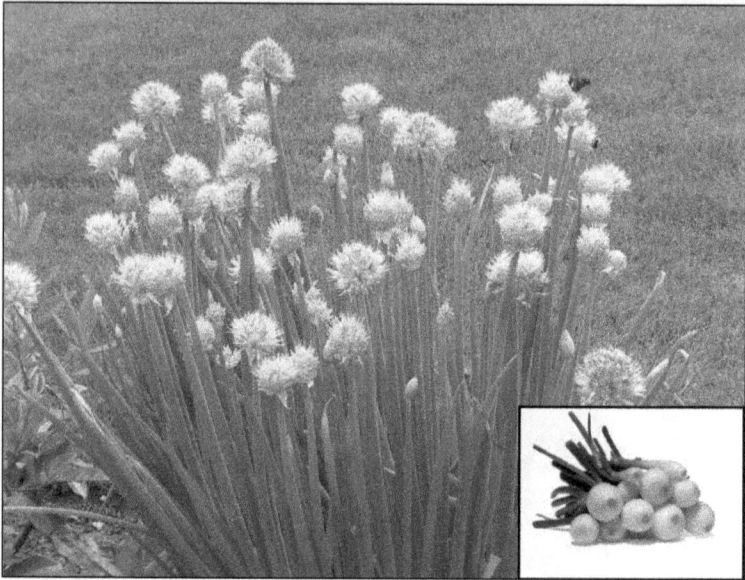

Allium fistulosum **Linn.**

alone or used as a flavouring agent in soups and stews. They are mild flavoured. The tender leafy varieties are raised for their edible tops while the others, for their white blanched leaf bases. Winter leek and stone leek are sometimes used as a substitute for each other.

Chemical Constituents

1. *Leek/Stone Leek*: Composition of stone leek is most similar to that of leek (*A. porrum*) as: protein, fats, minerals, carbohydrates, calcium, phosphorus, sulphur, iron, vitamin A, Vitamins B_1 and C.

2. *Leek Essential Oil*: New 85 volatile compounds are isolated from leek essential oil. Compounds identified as: thiols, mono-, di-, tri-, tetra- and pentasulphides, alkylthi-carboxylic acid, S-alkyl esters, thiophenes, trithiolsanes, other S-containing compounds, acetals, alcohols, aldehydes, ketones, esters, acids, furans, other oxygen-containing compounds and hydrocarbons.

37. Long Pepper (Pipli)
Piper longum Linn.

Piperaceae

Long pepper is indigenous to India. It is the dried fruit. The plant is a slender, aromatic climber with perennial woody roots. It occurs in the hotter parts of India from Central Himalayas to Assam, Khasi and Mikir Hills, Lower hills of Bengal, and evergreen forests of the Western Ghats from Konkan to Travancore (Kerala). It has also been recorded from Nicobar Islands. The plant has creeping jointed stems. Leaves are 5-9 cm long, 3.5 cm wide, ovate and cordate. Female spikes are cylindrical, pedunculate. Male spikes are larger and slender. Female spikes are 1.3-2.5 cm long and are 4.5 mm across. Fruits are ovoid, yellowish orange and minute drupes are sunk in fleshy spike. The female spikes are sharper, sessile and blunt at the tips and erect. These are red when ripe, flowers being numerous and packed.

It is mostly obtained from the wild plants. They are mostly supplied from Asam, West Bengal, Nepal and Uttar Pradesh, Kerala, West Bengal and certain parts of Andhra Pradesh. It is reported to be cultivated at lower elevations in the Anamalai Hills in Tamil Nadu and Cherrapunji area of Assam. It is cultivated in Meghalaya.

The long pepper as sold in India, appears to be derived from other species as well including one which is Indonesian. Indian

long peppers is a product of either of *P. longum* or *P. peepuloides*, while the Indonesian or Java long pepper imported from Malayasia, is from *P. retrofractum*. The fruits of these species are used for the same purpose and effectiveness.

Useful Parts
Fruits.

Chemical Constituents
Alkaloids, piperine and piplartine and two new liquid alkloids, resin, volatile oil, essential oil.

Processed Products
Volatile Oil
Dried fruits of long pepper on steam-distillation give 0.7 per cent of an essential oil, spicy odour, resembling that of pepper and ginger oils

Propertis and Uses
The fruits are used as a spice. These are also used in pickles and preserves. They have a pungent pepper-like taste and produce salivatin and numbness of the mouth when eaten or munched.

In Medicine
It is highly beneficial in indigestion, dyspesia flatulence, colic and cholera. The fruits and the roots are used for diseases of

respirator tract, *viz.* cough, bronchitis, asthma, etc. It acts as counterirritants and as analgesic when applied locally for muscular pains and inflammation. It acts as a sedative in insomnia, convulsions, hysteria and epilepsy. It is useful as cholagogue in obstruction of bile duct and gall bladder, bsides as an emmenagogue and abortifacient. It is beneficial in dysentery and leprosy. It promotes menstruation and regulates menstrual periods. It is efficacious to check haemorrhage and fever after childbrith. It benefits in gonorhoea piles and abdominal enlargements.

38. Marjoram (Sweet Marjoram)
Majorana hortensis Moench.

Lamiaceae (Labiatae)

It is a perennial aromatic herb. It is indigenous to South Europe, North America and Asia Minor. It is the dried leaves of marjoram or without flowering tops in small proportion that constitute the leafy spice of commerce. It is 30-60 cm high. It is extensively cultivated in India. Sweet marjoram is characterized by a strong spicy and pleasant odour. The flavour is fragrant, spicy, slightly sharp bitterish and camphoraceous. It grows in western Asia, South and North America, France, Germany, Hungary, Greece, Romania, Spain,

Portugal, U.S.A, England, the Mediterranean and North Africa. The dried herb is light green with a slight greyish tint. The leaves are grey-green, small on both sides. Many dot-sized oil glands are present on the leaf. They yield 3.5 per cent volatile oil.

Useful Parts
Leaves

Chemical Constituents
Protein, fixed oil, volatile oil, pentosans, tannin, ursolic acid, phosphorus, iron, silica, calcium, phosphorus, magnesium, manganese and chlorine.

Processed Products
1. *Volatile Oil*: Steam-distillation of the leaves and flowering heads yield a volatile oil, known in the trade as 'Oil of Sweet Marjoram'. The oil is colourless or pale yellow to yellowish-green, with a tenacious odour, reminiscent of nutmeg and mint. The oil of Indian origin consists of carvacrol, eugenol, chavicol, d-linlool, methylchavicol, d-terpineol and caryophyllene. Oils of European origin differ considerably from the Indian oil and have usually. terpenes (mainly terpinene) but are free from phenols; d-terpineol and terpineol are also present.

Properties and Uses
Leaves are used by the manufacturers for flavouring Polish sausages and cheese and also in soups, stews, dressings, salads, egg and vegetable dishes, cheese, fancy meat (lamb and mutton) dishes and sausages and poultry dressing. The leaves are used fresh or dried and are highly reputed as a condiment for seasoning foods. They are used also as a poultry-seasoner. Fresh leaves are employed as garnish and incorporated in salads. They are also used for flavouring vinegar. Dried flowering tops are used for sachets and potpourri. The aromatic seeds are used in confectionery and French confitures. The oil is employed to a small extent in high grade flavour preparations and perfumes, and in soap and liquor industries.

In Medicine
It is carminative, expectorant and tonic. Leaves and seeds are astringent. An infusion of the plant is used as stimulant, sudorific,

emmenagogue and galalactagogue. It is useful in asthma, hysteria and paralysis. Its oil is used as an external application for sprains, bruises, stiff and paralytic limbs and tooth-ache, and for hot fomentation in acute diarrhoea. Leaves and seeds of marjoram are astringent and are reported to provide a remedy for colic.

39. Mustard

1. *Brassica nigra* (Linn.) Koch
2. *B. alba* Boiss.
3. *B. juncea* (Linn.) Czern. and Coss. ex Coss.

Brassicaceae (Cruciferae)

Mustard is one of our most ancient spices. Pythagoras, the Greek Mathematician, wrote of mustard's medicinal properties five centuries before Christ. It is believed to have been widely used in Africa and China, centuries before Jesus. The early Romans liked to mix the sweet 'must' of new wine with certain crushed seeds which they called 'sinapis'. The resulting paste was called 'mustum arden' (hot must). Mustum arden became 'mustard' and the seed itself was named. Similarly, the Romans spread their fondness for mustard to Gaul and Britain. Preparations of mustard paste became hence more sophisticated. The town of Dijon (France) is said to have begun making its 'prepared mustard'. It then became internationally famous as early as the thirteenth century.

A 3,000 Worder by Alexander Dumas, for the 'mustard house', of Bornibus in Paris in 1870 was the longest food advertisement in history. Mustards many claims became famous. A colourful treatise on the history of 'prepared mustard' was reproducd in the 'Dictionary of Cuisine'. Of course, the modern era for mustard seed, actually began in 1720 when Mrs. Clements of Durham (England), found a way of milling the heart of the seed to a fine flour. It was industrialized in England in the nineteenth century. It became the standard method of processing the seed for use as a spice in cooking and also for the 'prepared mustards'.

1. *Yellow or white Mustard Seeds*: Three species of *Brassica* of economic significance are: (i) *B. nigra* (Linn.) Koch, provides black or "Trieste" mustard seeds (ii) *B.hirta* Moenchi or *B. alba* Boiss. white or yellow mustard seeds and (iii) *B. juncea* (Linn.) Czern. Coss. ex Coss. "brown and

Brassica nigra (Linn.) Koch

oriental mustard seeds" (also called "Indian mustard seeds"). The seeds of *Brassica hirta* (*Sinapis alba*) are relatively larger than the other types used in food. Generally, they are about 3 mm in diameter and flattened laterally. As the name above implies, it is mostly of a pale-straw colour. This may also have a slightly pinkish cast and there can be an occasional light brown one. The yellow mustard seed is sharp in taste due to a non-volatile flavouring substance. This gives it a sharp tongue taste but not aromatic pungency. Yellow mustard is thought to have originated in Europe. Today, it is widely cultivated there and the USA.

2. *Brown and Oriental Mustard Seeds*: These both types arose are from *Brassica juncea*. They have the same chemical composition, though the average fixed oil tends to be higher

B. alba Boiss.

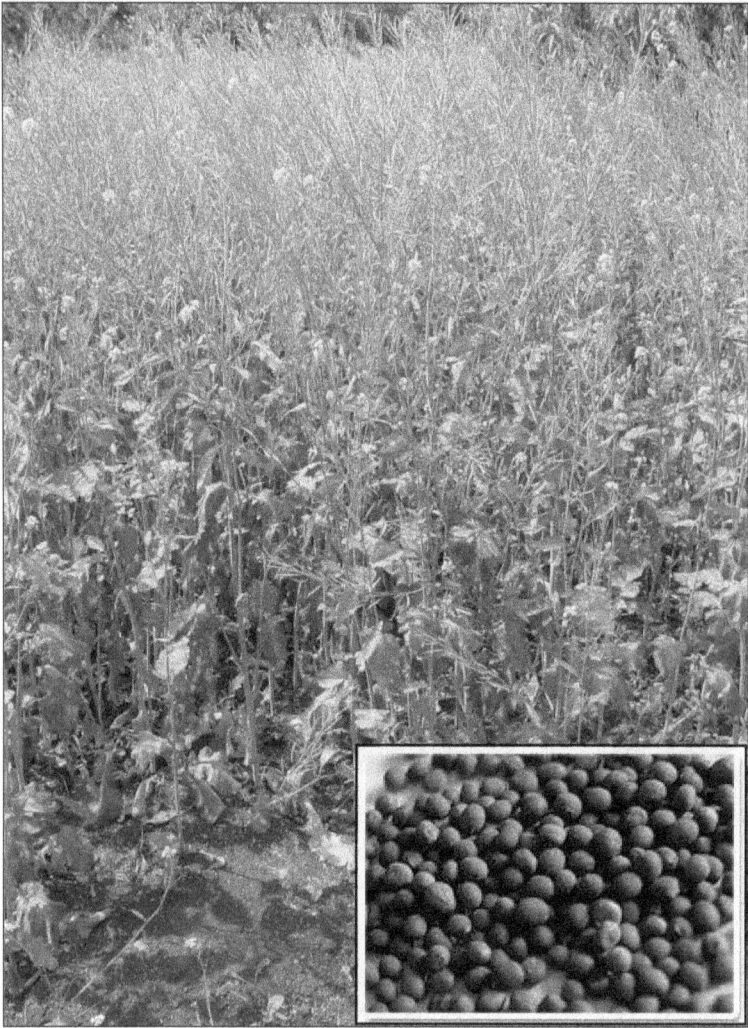

B. juncea (Linn.) Czern. and Coss. ex Coss.

in the oriental, while protein and fiber are typically higher
in the brown. The seeds of both measure about 2 mm in
diameter. The seeds of brown mustard are reddish-brown
to dark-brown, but the oriental seeds are mostly light
yellow occasionaly is brownish. The important difference
in the species (applying to both brown and oriental is that

the seed contains a volatile oil which produce a very pungent aroma as well as bite. Chinese restaurant mustard, the hot English mustard and Dijon and German mustards all exhibit this effect because all are made from brown or oriental seeds. The *Brasica juncea* types probably came from Africa, India and the interiors of China. They are widely cultivated in Canada and northern plains of the USA.

3. *Black Mustard Seeds*: The *Brassica nigra* seeds are about 2 mm or less in size. This species is considered a native to Europe and the Middle East and until comparatively recent times, it was the main seed of commerce in Europe. Since World War II however, browns and orientals have gradually replaced it because they can be grown more economically.

Useful Parts

Seeds.

Chemical Constituents

Volatile oil (allyl isothiocyanate, crotonyl, isothiocyanate, acrynil isothiocyanate) essential oil, fixed oil.

Procesed Products

The following are major processed products from condiment mustards and their admixtures in controlled conditions:

1. *Mustard Flour or Powder*: It is used in the manufacture of mayonnaise. It is also used as a condiment in pickles, meat and salad dressings. Indian black mustard is familiar for its flavour and pungency. A improved process for the production of 'Mustard Flour or Powder' from mustard seeds has been derived. This powder can be used in the preparation of 'mustard paste', 'table mustard' and other mustard products.

2. *Maonnaise*: It is one of the most important applications of the emulsifying qualities of mustard. It consists essentially of a creamy emulsion of vegetable oil in vinegar. The emulsifying agents present are mustard and egg yolk. Other substances are added for flavouring purposes without affecting stability. It is proved that mustard not only is useful for its condimental qualities in maonnaise, but in

addition, exerts a valuable emulsifying effect, which helps in the promotion of stability in mayonnaise.

3. *Dried/Dehydrated Mustard Greens (Leaves)*: Leaves are dried or dehydrated. Some of the Japanese mustard plants have large leaves and are exclusively grown for vegetables. These are also, at times, dried/dehydrated, packed and sold.

4. *Mustard for Industrial Buyer*:   The spice mustard comes in three forms for the industrial buyer :

   (*a*) *Whole Mustard Seeds*: The whole seeds are select, bold seeds of yellow type. They are used primarily in pickles, relishes and condiments, where appearance of whole seeds is an attraction.

   (*b*) *Powdered Mustard or Mustard Flour*: It is the most important processed form of mustard seed. It is often termed 'flour' which is one of its designations in the FDA standards. The husk (bran) of the seed is removed by milling and the heart ('mildings') of the seeds are finely ground. The heart meat is milled to a uniformly fine powder which disperses and blends particularly well in cooked foods, sauce products and 'prepared mustards'. 'Mustard flour' is usually a blend of oriental/ brown and yellow seed. The miller thus produces a product which can be adjusted to meet the buyer's end-use requirements.

   (*c*) *Ground Mustard:* It is also known as 'mustard meal'. It is the product of grinding the whole seed, husk included. It is a blend of yellow and brown/oriental seeds but is mostly prepared from yellow mustard alone. It is mainly used in the sausage industry. The point is that the husks contain a mucilaginous substance which acts as a good binding agent in sausage products. The ground mustard thus especially affords this feature in addition to flavour.

   (*d*) *Mustard Cake*: After removal of fixed oil wholly or partially from ground mustard, the resulting byproduct is termed 'mustard cake'. Since the oil removed is a vegetable oil which does not contribute to flavour, the other valuable properties remain the same in the cake.

Properties and Uses

Mustard seed is an important condiment used to flavour various food products. Whole mustard is used as a flavouring material after frying, in curries, pickles and sauces. It is used as a garnish in salads. Mustard flour is widely used as a condiment to impart flavour and piquancy to all kinds of meats. It is also employed in the manufacture of proprietory mustard powders such as 'Coleman mustard'. It is also used in 'prepared mustard' a paste product containing mustard flour, vinegar, spices, etc. It is employed in the well-known product named as 'mayonnaise', which consists of mustard and egg as emulsifiers, added to a creamy emulsion of edible oil and vinegar containing some other flavourants. Seed oil is used as hair oil in North India. The leaves are used as vegetable.

In Medicine

Mustard is rubefacient. Its paste is applied as an analgesic in rheumatism, sciatica, paralysis of limbs and muscular pains. Mustards seeds are recognized as a decongestant and an expectorant. They are effective against congestion caused by colds and sinus problems. Seeds are emetic and hence advised in narcotic, drunkenness and other poisonings. Seed powder is very effective in painful menstruation. It is beneficial in the treatment of convulsions in children. Seed paste is valuable in ringworm.

40. Myrtle
Myrtus communis L.

Myrtaceae

In ancient Greece and Rome, Myrtle was considered sacred and a symbol of youth, beauty and marriage. The sacred myrtle considered to be a phallic emblem, was held sacred to Venus, and among the Greeks and Romans. It was believed that it was behind a myrtle tree in the island of Cytherea that Venus took refuge from the Satyrs who disturbed her when she was bathing. She was wreathed in myrtle when Paris awarded the golden apple. And it was with myrtle boughs that she ordered the chastisement of Psyche for daring to complete with her beauty. The myrtle tree, which grows best by the sea from which Venus arose. It is specially dedicated to goddess Venus. Wreaths of leaves were worn by the Athenian magistrates, by the victors in the Olympic games and by others.

Much economic significance is also attached to this sacred tree in ancient times. The leaves were often pulverized by the people of Orient for use as an astringent dusting power for infants when in 'Swaddling clothes'. Myrtle powder was applied after the body had been rubbed down with olive oil to impart a pleasant fragrance to the child. The berries were employed as a spice by the Romans (but were little used since the discovery of black pepper). The Tuscans still prepare a wine from the berries, called Myrtidanum. Myrtle leaves, branches and berries have been used since Biblical times to flavour smoked or roasted meats in the Eastern Mediterranean islands of Sardinia, Corsica, Crete and Western Asia. Myrtle is an evergreen tree with leaves ovate – lanceolate, aromatic, white fragrant flowers, berries ellipsoid, blue-black, and kidney-shaped seeds.

Useful Parts

Leaves, flowering tops, fruits (berries).

Chemical Constituents

Leaves, flowering tops and berries: Essential oil (myrtenol, myrtenol acetate, a-pinene, limonene, linalool, camphene, cineol, geraniol, nerol); berries also contain citric acid, malic acid, resin and tannin; seeds contain fatty oil (glycerides of oleic, linoleic, myristic and palmitic acids).

Properties and Uses

The leaves are astringent and bitter with a refreshing fragrant and orange like aroma. Berries are sweet with juniper and rosemary like flavours. The berries are used in deserts, liqueurs, meats and meat ragouts. The leaves are used to wrap wild jam meat or roast pork before cooking. Myrtle oil is used as a substitute for dried leaves as a flavouring agent for culinary purpose. It is used for scenting soups and toiled waters.

Processed Products

Myrtle oil is obtained on steam distillation using fresh leaves and flowering tops. This oil is yellow to greenish-yellow with a characteristic aroma. It is used as a flavoring agent.

In Medicine

The berries are useful as an antiseptic for bruises. The leaves are used to relieve gingivitis and sinusitis. The leaf powder or infusion is useful wounds, eczema and ulcers.

41. Nutmeg and Mace
Myristica fragrans Hout.

Myristicaceae

Nutmeg and Mace are two distinctly different spices produced by the same fruit. The fruits are borne by a green aromatic tree, which reaches to 20 m. height. Mace is the dried reticulated 'aril' of the fruit. Nutmeg is the dried seed-kernel of the same fruit. The mace when the fruit opens is seen as an attractive, bright, scarlet-cage, closely enveloping the hard, thin, black shining shell of the seed, called 'nutmeg'. At this point, the fruits is harvested with a special device. The husk (rind) is broken apart carefully. The mace is skillfully removed from the seed shell. It is then gently pressed flat and dried in partial shade. The typical 'blade of mace' is scarlet in colour. After drying it turns rather pale yellowish-brown or reddish-brown and becomes brittle. The husk is generally pickled or candied. The nutmegs (kernels) are left within the shell and dried until they rattle when shaken. The very thin shells are then carefully broken. The kernels (real nutmeg) are removed, air-dried and packed.

The flavour of mace is more than that of nutmeg. Nuts are ovoid, approximately 2.25 to 2.75 cm long, 1.75 to 2.25 cm in diameter, and longitudinally wrinkled. The colour is greyish brown, with furrows and network of dark-brown veins, in which volatile oil is found. *M. fragrans* is a native of Moluccas, now cultivated in many tropical

countries of both hemispheres *viz.*, Malasia, Indonesia, West Indies, Grenada, Sri Lanka, India, etc. In India, it is grown on a small scale in Tamil Nadu (Nilgiris, Burliar, Coimbatore Salem, Ramanathapuram, Tirunelveli, Kanyakumari and Madurai), Kerala, Assam and other states.

Useful Parts

Fruit (seed), Aril.

Chemical Constituents

1. *Nutmeg*: Protein, carbohydrates, mineral matter, volatile oil, starch, pentosans, furfural, and pectin.

2. *Mace*: Protein, carbohydrates, mineral matter (Ca, P, Fe), volatile oil, amylodextrin, reducing sugar, pectin, resins, saponin, vit. A, vitamins B_1, B_2, niacin and vit. C, volatile oil (amylodextrin).

Processed Products

(A) Nutmeg Products

1. *Volatile Oil*: The oil is derived from broken and wormy nutmegs. The material is comminuted, pressed to remove

fixed oil, and immediately subjected to steam distillation. Loss of volatile oil from ground nutmegs is relatively rapid (in about 2 months). Cohobation of distilled waters may be necessary for the recovery of the total oil. Oil of nutmeg is a mobile, almost colourless or pale-yellow liquid with a characteristic odour on ageing. It partly resinifies and becomes viscous on storage. The aroma of the East Indian oils is much more pronounced and more characteristic of the spice than that of West Indian oils.

2. *Nutmeg Butter*: Nutmeg contains 38-43 per cent of ether extractable material which, in addition to glycerides, contains a volatile oil (6-13 per cent), a small quantity of resin, and a substantial proportion of unsaponifiable material. Commercial nutmeg butter, a highly aromatic fat, is obtained from undersized, damaged or worm-eaten kernels which are unfit for sale as spice. The material is ground and cooked or steamed before pressing. The fat may be obtained by solvent extraction also but this process is not usually employed. Nutmeg butter is soft, solid, yellow or yellowish-red, with the odour and taste of nutmeg.

(B) Mace Products

1. *Oil of Mace*: It resembles nutmeg oil in odour, flavour and composition and no distinction is made between them in the trade. Like nutmeg oil, mace oil also becomes viscous on storage due to absorption of oxygen. Old mace yields more viscous oil than the fresh one.

2. *Fixed Mace Oil*: Mace yields a fat similar to that from nutmeg but in a much lesser quantity. The amylodextrin is present in mace in the form of granules. It is visible under the microscope (5-7m). They are compound and irregular in shape with a distinct hilum.

Properties and Uses

Both nutmeg and mace are used as condiment, particular in liquor, sweet foods, baking, etc. Oil of nutmeg or mace is employed for flavouring food products and liquor. It is used for scenting soaps, tobacco and dental creams, and also in perfumery.

Nutmeg butter is used as a mild external stimulant in ointments, hair lotions and plasters. It is used in perfumes for imparting a spicy odour and in the manufacture of soaps and candles.

Alcoholic extracts of nutmeg show antibacterial activity against *Micrococcus* var. *aureus*. Aqueous decoctions are toxic to cockroaches. Myristicin present in the kernel may be employed as an additive to pyrethrum to enhance toxicity of the later to houseflies; myristicin by itself is inactive. The volatile oil from the leaf has weedicidal properties. It may also be used for scenting soaps, dentifrices, chewing-gums and tobacco. It is not produced on a commercial scale.

Nutmeg and mace are generally classified as baking spices since both are particularly good in sweet foods. However, they find a much wider range of use than other baking spices. They are frequently included in frankfurter formulae and in recipes for curried meat and other products. They are much used in soups and preserves, pound cakes, cream picks and soaps in sauces in combinations with dairy products. In India, mace and nutmeg are used more as drugs.

In Medicine

In eastern countries, they are used more as a drug than as a condiment. Nutmeg is a stimulant, carminative, astringent and aphrodisiac. It is used in tonics and electuaries and forms a constituent of preparations prescribed for dysentery, stomach-ache, flatulence, nausea, vomiting, malaria, rheumatism, paralysis, sciatica and early stages of leprosy. Excessive doses have a narcotic effect; symptoms of delirium and epileptic convulsions appear after hours. Mace is similarly used. It is also chewed for masking foul breath. Oil of the nuts is used in inflammations of bladder and urinary tract. Both are stimulant, carminative, astringent and aphrodisiac and are used in pharmaceutical preparations for dysentery, stomach-ache, flatulence, nausea, vomiting, malaria, rheumatism, sciatica and leprosy (early stage).

42. Onion
Allium cepa Linn.

Liliaceae

Onion (also garlic and leak) as food, medicine and religious objects can be traced back to the 1st Egyptian Dynasty (3200BC) and

also to the biblical account of the exodus of the Isralites. Their use in medicine in India can be inferred form Indian writings. By the time of the Greek and Roman authors, from Hippocrates (430 B.C.).Theophrathus (322BC.) to Pliny (79A.D.) several cultivars were named. They are described as mild pleasant, round, long white, yellow, red, etc. Thus onions are mentioned in the literature from Hippocrates (430 BC.). Onions were depicted as food in Egyptian tombs as early as 3200 BC. It was consumed by the builders of the pyramids and was used in religious and funerary offerings. It has been found in mummies. It is also mentioned in the Bible (xi,5) and in the Koran. It is mentioned in Garuda Purana as 'Palandu'. Maharishi Atrey and Lord Dhanvantari also described its uses. Onion juce is mentioned to improve sexual power in 'The Perfumed Garden", a 16th century Arabic erotic manual written by Sheikh Al Nafzawai. It is used both for cooking as a food and as a condiment

for flavouring or for pickling. India is one of the world's largest producers of onion and has been exporting onions to South-East Asian countries. It is grown almost all over India. The most important onion-growing states are: Maharashtra, Tamil Nadu, Gujarat, Uttar Pradesh, Andhra Pradesh, Odisha and Bihar. The US is the largest producer of onions, followed by Japan, Spain and Egypt. Onion is herbaceous annual for bulb production. But for seed production it is biannual.

Various types of Onion are grown in India useful for particular uses for *e.g.* (*a*) Green onions are used in salads. (*b*) Mild onions are used for cooking as food or as salad. (*c*) Pungent varieties are used as condiment for flavouring of a number of foods. (*d*) Pearl onions or small onions are used in pickles, including vinegar pickles (*e*) White and yellow onions for dehydration purpose and for manufacturing onion powder; white onions of the desired quality are preferred and (*f*) Red and yellow onions are general purpose crops for culinary use.

Processed Products

On a commercial scale, a number of products are manufactured from white onions, wherein dehydration is involved, for example dehydrated onion, onion flakes, onion rings, kibbled onions, onion powder and onion salt. Small pearl onions are used 'whole' in vinegar pickles. Recently onion paste is also being manufactured for use in curries, etc.

1. *Dehydration of Onion*: It is produced by removal of water from the raw onions to a maximum level of 4.25 per cent and them milling it to a specific particle size. Onions are dehydrated without blanching or sulfiting in order to protect the enzyme system which develops onion flavour when onion cells are cut or broken. When the cells are broken, the onion enzyme alliinase is free to contact certain onion compounds which reacts to form numerous volatiles; responsible for the charactristic onion flavour.

2. *Onion Powder*: It is prepared by grinding dehydrated onion slices in a hammer mill to a suitable mesh. It is highly hygroscopic and hence the important precaution regarding its storage is to keep it in air-tight containers in a cool, dark and dry place.

3. *Onion Salt*: It is prepared by mixing 19-20 per cent of onion powder with 78 per cent free-flowing pulverized refined iodized table-salt and 1-2 per cent anti-caking agent like anhydrous sodium sulphate. This prevents water absorption and caking of the onion powder during storage.

Useful Parts
Bulbs, Leaves.

Chemical Constituents
Onion powder contains moisture, protein, fats, carbohydrates: mineral matter (calcium, phosphrus, sodium, potassium, iron), vit. A, vit B_1, vit B_2, nicotinic acid. Onion bulbs contain proein, fat, carbohydrates, essential oils, chief constituent of the crude oil is allyl-propyl disulphide. Vit. C., carotene, thiamine, riboflavin.

Properties and Uses
Onions are used as salad, and are cooked in several ways in all types of curries. They are baked, fried and used in fresh or dehydrated or powder form in soups, pickles, sauces, etc. They are also eaten raw as salad which helps in lowering blood pressure and blood cholesterol. Onion oil, produced by steam distillation is used for flavouring foods.

In Medicine

The leaves are diuretic, carminative, digestive, emollient, tonic, alternative, anthematic, stimulant, expectorant, antispadomodic, mild laxative and aphrodisiac. Onion has antiseptic properties due to the presence of several sulphur compounds. Onion is stimulant, diuretic and expectorant. It is considered useful in flatulence and dysentery. Freshly expressed onion juice has moderate bactericidal properties.

Onion (bulb) is a mucus clearing food. It has been in use for centuries in cold, cough, bronchitis and influenza. It is preventive against heart attack. A syrup made from onion and honey has been found very effective in restoring sexual power. Warts of skin disappear when rubbed with cut onions, Poultice of roasted onions is applied against indolent boils, bruises and wounds. Onions are beneficial in diabetes. They are effective in cholera, ear infection, rheumatic diseases, aches and pains, urinary disorders like burning secretion with micturition and dsymenorrhoea.

43. Origanum
Origanu vulgare Linn.

Lamiaceae (Labiatae)

It is popular by differant names Origanum, Oregano, Mexican Origanum or Sage the Pizza Herb Origany or Wild Marjoram. It is one of the most popular herbs in Mediteranean cooking. It constitutes the dried leaves of an aromatic, branched, perennial herb, 30-90 cm high. It has a creeping root system and a rather sprawling habit. The plant is used as a pot-herb or vegetable. Leaves are broadly ovate, entire or rarely toothed, about 1.5 cm. long, and light green when dried. Flowers are purple or pink in corymbose cymes. Nutlets are smooth and brown. The colour of the dried herb is light green. The aroma of the herb is strong camphoraceous and resembles to that of marjoram. In fact, the Spanish work 'oregano' means 'marjoram'. The taste is fragrant, spicy, warm, pungent and bitter. The plant owes its usefulness as a culinary herb to its volatile oil. The herb is traded both as 'whole' dried leaves and in 'ground' form. Oregano is cultivated in Greece, Dominican Republic, Turkey, Sicily and Italy. It grows abundantly in Mexico. It is very common in Shimla Hills and Kashmir Valley. Oregano is dried when in flowering in order to allow full development of its aroma or flavour. Stalks may be picked

as they blossom, and hung up in small bunches in shade for drying. The dried leaf is about 1.5 cm long and light green in colour. Generally, it is available as rubbed (or crushed) or ground form. It is an important pizza spice. Its quality is measured by its volatile oil content, moisture, total and acid-insoluble ash.

Useful Parts
Leaves.

Chemical Constituents
Protein, fat, carbohydrates, minerals, calcium, phosphorus, iron, sodium, potassium, vit. A, vit B_1, vit B_2, niacin, and vit. C.

Processed Products
(i) Volatile Oil
Origanum contains volatile oil, fixed oil, cellulose, pigment and mineral elements. Oil of origanum appears to be sold under this name from a number of related species of Labiatae plants. In some, origanum oil, thymol is the main constituent; in others, it is carvacrol. Botanists and herb-growers classify it as the pizza herb.

The herb contains a volatile oil, tannin and a bitter principle. The oil of European origin possesses an aromatic spice, somewhat basil-like odour and contains thymol, carvacrol, free alcohols, esters as geranyl acetate and bicyclic sesquiterpene. It is freely soluble in 90 per cent alcohol; contains 1-pinene, dipentene, linalool, bi- and tri-cyclic sesquiterpenes and palmitic acid.

Properties and Uses

No Mexican kitchen is without oregano since they believe that no other herb imparts or communicates such an excellent aroma and flavour to food. The leaves and tops, prior to blooming, are used to flavour foods in the same way as sweet marjoram (*Majorana hortensis*). Oregano is an essential ingredient of chilli powder. It is used in 'chilli concarne' and many other Mexican dishes. It is the spice that made pizza famous and is equally good in any tomato-type dish, from spaghetti to old-fashioned stewed tomatoes. Origanum is also used for flavouring soups, meat-dishes, pork, fish, egg-dishes and salads. The plant is used in Punjab as a pot-herb. It is eaten also as vegetable in Lahaul. The oil is used in the cosmetics and soap industries.

In Medicine

The oil has carminative, stomachic, diuretic, diaphoretic and emmenagogue properties. It is administred as a stimulant and tonic in colic and diarrhea. It is also used in chronic rheumatism, tooth-ache and ear-ache. It is used in whooping cough and bronchitis. In homeopathy, it is used for hysteric condition. It is used as an external application in healing-lotions for wounds, usually in conjunction with other herbs. The oil has been employed in veterinary liniments. It is used in gargle and bath. It stimulates growth of hair.

44. Parsley
Petroselinum crispum (P. Miller)
Nymann ex A.W. Hill
Apiaceae (Umbelliferae)

It is a native to Sardinia. It is cultivated in the Mediterranean region and the USA. It is a hard, aromatic biennial, much-branched green herb, sometimes lasting up to four years, producing a rosette of finely divided radical leaves in the first ear. It bears a flowering stalk up to 100 cm high in the second ear. It has rich green compound

leaves 2- or 3- pinnate. Fruit is 2-3 mm long, crescent-shaped, conspicuously ridged, consisting of two mericarps. Leaves and seeds are used as spice. The colour of the dried herb is green. Its aroma is pleasant, characteristic, fragrant and spicy due to volatile oil present therein.

There are two main types o horticultural parsles: (i) the one cultivated for the leaves (var. *crispum*) and (ii) the other grown for its turnip-like roots (var.*redicusum* Danert). Only the former type of parsle is cultivated in India. In the latter case, roots are cut after the fruits (seeds) are harvested.

Useful Parts
    Leaves, Roots.

Chemical Constituents
    Protein, fat, carbohydrates, minerals, (calcium, phosphorus, iron), carotene (as vitamin A), thiamine, nicotinic acid, vit. C (ascorbic acid), riboflavin.

The leaves, stems and fruits contain a glucoside 'apiin', which on hydrolysis, yields 'apigenin', glucose and a sugar, apiose; a

second glucoside, consisting of luteolin, glucose and apiose has also been reported.

Processed Products

1. *Dehydrated Parsley*: The whole dried parsley emerges onto a mechanical and air-separating system, where the stems are removed from the leaves. The leaves are sold as 'flakes' or 'granules'; leaves and stems are ground to powder. The entire process of parsley dehydration as described above, from field to final packages, is accomplished in less than 02 hours. In today's quick modern dehydration process, 08 parts of the fresh parsley are reduced to just 01 part or its dehydration ratio is 8:1.

2. *Volatile Oil*: Entire plant contains essential oil, well known as 'Oil of Parsley'. It has the characteristic aroma and

flavour of parsley. The oil is recovered by steam-distillation and is used mainly for flavouring food products. The oil obtained from the flowering-tops is of the finest quality. Commercial parsle oil is distilled either from the aerial parts of the herb, bearing immature fruits or from the mature fruits. The herb oil has superior aroma and is more reputed than the fruit oil. The fruit (seed) oil contains apiol (parsley camphor) and a- pinene, with small amounts of myristicin, aldehydes, ketones and phenols. The herb oil is reported to contain 'apiol'. The apiol is used medicinally for the same purposes as the herb.

3. *Fatty Oil*: The fruits yield a greenish fatty oil with a peculiar odour and disagreeable sharp flavour. The oil has a high content of petroselinic acid (up to 76 per cent). It is useful for a variety of industrial purposes *e.g.* making plastics, synthetic rubber, lubricating oil additives and protective coatings, etc.

Properties and Uses

Fresh leaves are generally used for garnishing and seasoning of foods. They are eaten fresh used in salads and as an ingredient of soups, stews and sauces. They are also used in meat and poultry seasoning. The leaves are used to prepare a tea which is thought to possesss antiscorbutic properties, being very rich source of vitamin C. The roots are used as a vegetable in soups. The dried leaves and roots are used as condiments.

In Medicine

The herb has diuretic carminative, ecbolic, emmenagogue and antipyretic properties. It is useful for uterine troubles. The juice of the fresh leaves is used as an insecticide. Bruised leaves are applied to bites and stings of insects, and the mericarps are used to get rid of lice and skin parasites.

45. Peppermint
Mentha piperita Linn.

Lamiaceae (Labiatae)

Peppermint is a perennial, glabrous, strongly scented herb grown or cultivated in temperate regions of Europe, Asia, etc. It is considered to be the hybrid between. *M. spicata and M. aquatica.* It is

sensitive to drought. It is erect, 30-90 cm high, purplish or green; leaves are ovate or oblong, coarsely serrate, smooth and dark green above.

Useful Part

Leaves.

Chemical Constituents

Methyl acetate, menthone.

Processed Products

1. *Volatile Oil*: The herb is the source of true peppermint oil. Commercial oils are derived from cured, partially dried plants, while official oils are obtained from fresh material. The yield of oil varies from 0.3 to 1 per cent depending on the extent to which the material has been dried before distillation. Peppermint oil is a colourless, pale-yellow or

greenish-yellow liquid with a strong agreeable odour and a powerful aromatic taste, followed by a cooling sensation when air is drawn into the mouth. On ageing, the oil darkens in colour and becomes viscous. When chilled, menthol separates out as crystals. The oil contains menthol (50-55 per cent), methyl acetate, menthone and small amount of several other compounds.

Propertis and Uses

The green plant, left after the extraction of oil, may be dried into hay or silaged for use as a cattle-feed. The hay contains: protein, digestible protein and digestible nutrients. It may be employed as a substitute for Lucerne-hay for feeding dairy cows.

In Medicine

The herb is aromatic, stimulant, stomachic and carminative. It used for allaying nausea, flatulence and vomiting. Bruised leaves are used as an external application for relieving local pains and headache. A hot infusion allays stomach-ache and colic diarrhoea.

Peppermint oil is employed for flavouring pharmaceuticals, dental preparations, mouth washes, cough drops, soaps, chewing gums, candies, confectionery and alcoholic liquors. It is valued in medicine both for internal and external uses. For internal use, it is preferred to menthol because of its more pleasant taste. It is widely employed in flatulence, nausea and gastralgia. It may be administered with sugar or in the form of tablets and lozenges. The oil is mild antiseptic and anaesthetic. It is used as an external application in rheumatism, neuralgia, congestive headache and tooth-ache.

46. Poppy Seed
Papaver somniferum Linn.

Papaveraceae

Poppy probably originated in Turkey as a cultigens. Arab traders spread its habit and the plant westwards and it may have been used by the Swiss Lake Dwellers. The plant reached India and China by the 18th century A.D. Poppy seed is so tiny that one pod generally contains 30,000 seeds. About 9,000,000 are required to weigh a pound. Each seed has a tiny droplet of nut-sweet oil which explains its flavour appeal. Although poppy seed comes from the

same plant which produces opium. The seed is formed after the plant has matured and it has lost all of its opium potential. There is absolutely no narcotic content in poppy seed.

Poppy seed is grown in the Netherlands, Poland, Denmark, Sweden, the Balkan countries, Turkey, Argentina, and to some extent, in India. Poppy is cultivated either for manufacture of opium or for seeds. In India, var. *album* of poppy with white seed has been cultivated for many years for the production of seeds under license in some states. In Europe, var. *nigrum* which has slate-to blue-

coloured seeds, and known as 'Maw Seeds', is exclusively cultivated for this purpose.

Useful Parts

Seeds.

Chemical Constituents

Seed contains calcium, phosphorus, iron, thiamine, riboflavin, and nicotinic acid, carotene is absent. Minor minerals in the seeds include iodine, manganese, copper, magnesium, sodium, potassium and zinc. The seeds also contain lecithin, oxalic acid, pentosans, traces of narcotine and an amorphous alkaloid, and the enzymes diastase, lipase and nuclease.

The seeds have a high protein content, the major component being (globulin) which accounts for 55 per cent of the total nitrogen. The amino acid make-up of the globulin is similar to that of the whole seed protein and is as follows: arginine, histidine, lysine, tyrosine, tryptophan, phenylalanine, cystine, methionine, threonine and valine.

Processed Products

Poppy Seeds Oil

Poppy seeds contain up to 50 per cent of an edible oil which is extracted by either cold or hot expression. The oil is odourless and possesses a pleasant almond-like taste. The oil is generally extracted in India by cold pressing the seed in small presses in homes or small establishments. Raw cold-pressed oil is pale to golden yellow in colour. It is used in soap and medicines.

Properties and Uses

The poppy seeds are used as a source of fatty oil. They are considered nutritive and are used in breads, cakes, cookies and pastries, curries, sweets and confectionery. Poppy seeds or seed meal also find use in the production of lecithin.

Poppy-seed oil is consumed without refining. It does not develop rancidity easily. Hot-pressed oil is largely used in soap-making. It may be rendered edible by refining. The oil yield from black and white seeds is nearly the same. However, black seeds are more commonly used for expression of the oil because of its easier cultivation. Nevertheless, white seeds are reported to yield the finest oil. Seeds from the capsules which have not been scarified for opium give a higher yield of oil than from those scarified.

Poppy-seed is commonly used for culinary purposes. It is free from narcotic properties. It is used in combination with olive oil, or as a salad oil. On hydrogenation, it yields a product similar to hydrogenated groundnut oil which may also be useful for industrial purposes. Poppy-seed oil is used in the production of artists' paints; for this purpose sun-bleached oil from the first cold pressings of the seed is preferred. The oil is useful in the preparation of linoleic acid, soft soaps, ointments and emulsions, and compositions for skin care. The oil finds use also as an illuminant.

The cake or the meal left after extraction of the oil from the seeds is sweet and nutritious. It is eaten by poor people. It is readily consumed by cattle and sheep, and may be fed alone or preferably mixed with other feeds.

The capsules contain the same constituents as opium but in much smaller quantities. The capsules contain about 70 per cent of the total morphine of the plant. The morphine in the capsules decreases rapidly on storage. Narcotine content of the capsules is reported to be 0.1-0.2 per cent.

In Medicine

An infusion of capsules is used as a soothing application for bruises, inflammatory swelling and sometimes for painful conjunctivitis and inflammation of the ear. A hot decoction of capsules is applied as an anodyne. Capsules are also used in the form of syrup or extract, as a sedative against irritant coughing and sleeplessness. In India, an intoxicating liquor is prepared by heating capsules with jaggery and water. In the USA, mature poppy capsules have been processed to yield a liquor which can be used as a substitute for opium in the production of morphine. Seeds are demulcent and are used in the form of emulsion, as an emollient and specifically against obstinate constipation and in catarrh of the bladder. The white seeds are sometimes used in pharmaceuticals. It is used against diarrhoea, dysentery and scalds. Lower grades are used for lubrication. Leaves are also smeared as an anodyne. The red poppy flowers are used in medicine for making syrup.

Seeds are demulcent and used in the from of emulsion an emollient. Seeds are sedative and advised to treat dysentery. Rubbing of seed paste on body helps remove burning sensation, paste of seeds also beneficial in dry itch. Opium is useful in rheumatism, tumours, carbuncles, ulcers, leprosy, syphilis, scrofula etc. Opium is useful as a liniment for smoothing muscular and neuralgic pains. An infusion of the root is administered as a febrifuge. The essence of the root is used as tonic because of its stimulating qualities.

Poppy Straw

Dry, unlanced empty capsules alongwith a stem of about 7.5 cm are used in Europe and other places, where the plant is cultivated primarily for its seed and oil, as a source of morphine and narcotine. The two alkaloids are obtained in a yield of 0.08 and 0.009 respectively. Poppy straw and heads have also been processed to yield concentrated alkaloidal extract such as 'Optopon' (morphine: 20-22 per cent ; other opium alkaloids: 16-18 per cent) which can be used directly as a pharmaceutical. Poppy straw has also been tried for the manufacture of hand-made boards.

Young poppy plant is sometimes eaten like lettuce. It is grown as a pot-herb in Iran and is also fed to cattle.

Leaves and petals have been used for packing opium. The poppy plant is useful in the production of paper pulp. Kraft pulping gave pulps having good strength properties and with usefulness for the manufacture of wrapping papers, bags and other grades requiring an improved formation.

47. Rosemary
Rosmarinus officinalis Linn.

Lamiaceae (Labiatae)

Rosemary grows wild and is also cultivated in Yugoslavia, Spain, Portugal, France, and other parts of Europe as well as in California in the USA. It is a native of southern Europe and grows wild on dry rocky hills in the Mediterranean region. It has been suggested as suitable for cultivation in temperate Himalayas and Nilgiri Hills with dry to moderately moist climates. Rosemary comprises dried leaves derived from an exotic, leaf, evergreen shrub cultivated in Indian gardens in cool places for its pleasantly fragrant leaves. The leaves are narrow, about 2.5-cm long, and resemble curved pine needles. The dried herb is brownish green. Its leaves have a tea-like fragrance. The crushed rosemary, however, has an

agreeable and fragrant, spicy aroma with a camphoraceous note. The taste has fragrant, spicy, pungent, bitter and camphoraceous notes. Rosemary is marketed only in the whole form. Its name 'Rosemary' joins two Latin words meaning 'Dew of the Sea', because it thrives best where fog rolls in from the sea, as is the case along with its native Mediterranean region. The shrub bears a few bluish, white or violet flowers.

Useful Parts
Leaves, Roots.

Chemical Constituents
Protein, fat, carbohydrates, minerals (calcium, phosphorus, iron, sodium, potassium), Vit. A, Vit. B, niacin, saponin, tannin, ursolic acid, carnosic acid, amyrins, betulin and rosmarinic acids.

Processed Products

1. *Volatile Oil*: It is obtained by steam-distillation of leaves, flowering-tops and twigs. The bulk of the commercial oil comes from the plants growing wild in Spain; with smaller quantities from France, Dalmatian Islands, Tunisia and Morocco. Indian requirements are met with only by import. The characteristics of the oil vary with the source and also with the parts of the plant distilled; the finest product being obtained from the dried leaves, freed of stalks. The oil is pale yellow or almost colourless liquid with the characteristic odour of the leaves and a warm camphoraceous taste. The chief constituents of the oil are pinene, camphene, cineol, camphor, borneol and bornyl acetate.

Propertis and Uses

Fresh tender tops are used for garnishing and for flavouring cold drinks, pickles, soups and other foods. The leaves are used as a condiment, dried or powdered. These are added to cooked meats, fish poultry soups, stews, sauces, dressing, preserves and jams. They are mixed with sage in pork and veal stuffings, and also in biscuits. Oil of rosemary is used mainly in cheap perfumer, scenting of soaps, hair lotions and denaturing of alcohol. It is used in room-sprays and inhalants as well. Superior grades of the oil are employed for blending in 'eau-de-cologne' and for flavouring meats, sausages, soups, table-sauces and other food products.

In Medicine

Internally, the oil may be taken as a stimulant in doses of a few drops; a 5 per cent tincture is used as a circulator and cardiac stimulant. The oil is useful in headache and in tardy menstruation. It is diphoretic and is employed with hot water in chills and colds. An emulsion prepared from the oil is used as a gargle for sore-throat. The oil exhibits antibacterial and protistocidal activities.

Flowering-tops and leaves are carminative, diaphoretic, diuretic, aperient, emmenagogue, stimulant and stomachic. They are used for vapour baths for the relief of rheumatism, paralysis and incipient catarrhs. Dried leaves are smoked for the relief of asthma. A decoction of leaves is employed as an abortifacient. All parts are useful as a nervine tonic and an excellent stomachic. An infusion of the plant, with borax, is employed as a hair washes and prevents premature baldness. The plant has been found useful in atonic dyspepsia.

48. Saffron
Crocus sativus Linn.

Iridaceae

Saffron is the most valuable spice in the world. It is an important food flavourant and food colourant. It is also reputed to possess several medicinal properties. The recorded accounts of saffron cultivation in Kashmir date back to 550 A.D. Its recorded cultivation in Spain by the Arabs is around 961 A.D. The saffron fields, from mid-October to early November, when saffron plants are in full bloom, on a moonlight night, present a wonderful sight and a good source of fascination to many a tourists. Saffron consists of dried tri-lobed stigmas of *Crocus sativus*. It is a bulbous perennial with globular corms. The plant is only 15-25 cm high. It is native of southern Europe and cultivated in Mediterranean countries, particularly in Spain, Austria, France, Greece, England, Turkey and Persia, Iran, India (Jammu and Kashmir) and the Orient. Saffron is one of the oldest and certainly among the world's most expensive spices. One pound of saffron consists of about 2,25,000 to 5,00,000 dried stigmas and requires picking (by hand) of about 75,000 to 1,25,000 flowers. That

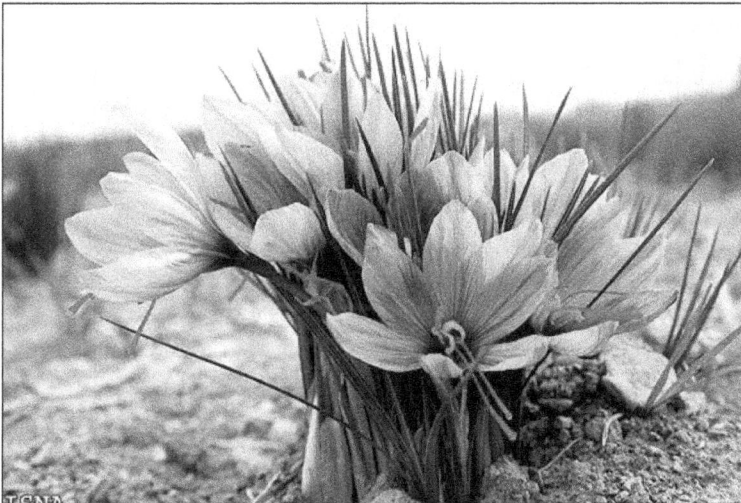

gives one a clear idea of the magnitude of human labour involved in the harvesting of saffron. The colour of the saffron is bright yellow-red or orange. The aroma is powerful, peculiar and exotic.

The plant is hysteranthous. It is a low-growing plant with an underground globular corm, and is cultivated for its large, scented, blue or lavender flowers. The flowers have trifid orange-coloured stigmas which, along with the style-tops, yield the 'saffron of commerce'. Pure saffron is plucked during early dawn before sunrise. This unusual way of plucking saffron flowers is practised till today in order to maintain freshness of the filament in the flower and to retain the colour of the saffron filament.

Useful Parts

Styles.

Chemical Constituents

Saffron, starch, sugars, essential oil, fixed oil, total N-free extract, minerals (potassium, phosphorus, boron), crocin, essential oil (terpenes, terpene, alcohols, esters), picrocrocin.

Propertis and Uses

Saffron is used extensively for rice dishes in Central Asia and Northern India. Indian sweets and saffron flavoured butter lassi are ever-lasting culinary impressions for everybody. Saffron is used also abroad in exotic dishes particularly in Spanish rice specialties and French fish preparations. It is also used in fine bread in Scandinavia and Balkans.

In Medicine

Saffron is useful to promote and regulate menstrual period. It benefits in leucorrhoea, hysteria and uterine sores. It is used occasionally in exanthematous diseases to promote eruptions. It is used in fevers, melancholia and enlargement of liver and spleen. It is a remedy for catarrhal affliction of children. In modern pharmacopoeias, it is employed only to colour other medicines or as a cordial adjunct. Saffron has been employed as an abortificient. It can be used as dressings for bruises and superficial sores.

In India, Saffron is an important ingredient of the Aurvedic and Unani systems of medicine. It is stimulant, warm and dry in action and helps in urinary, digestive and uterine troubles. It is mixed with other drugs to help in normal menstruation. If soaked overnight in water and used with honey, it enables patient suffering from urine trouble to pass urine freely. In homeopathy, the mother-tincture, prepared from saffron are useful in human disorders relating to mind, head, eyes, nose, throat, stomach, male and female organs, anus, respiratory and sleep problems. Its oil is used as an external application in uterine sores. When pounded, it is very effective in diabetes. It is also reported to give strength to brain.

49. Sage
Salvia officinalis Linn.

Lamiaceae (Labiatae)

Since ancient times in Europe, Sage has been used as a seasoning and medicine. Italians associated it with wiseness strengthening the memory and longevity. Sage was called 'Herba Sacra', meaning sacred herb.

Sage is the dried leaf of the subshrub. It grows and is also cultivated in Yugoslavia, Italy, Albania, Turkey, Portugal, Spain, Cyprus, England, Canada, the USA, and several European countries. It is a native of the Mediterranean area.

It is hard and variable sub-shrub, and is often cultivated as a spice and for medicinal purposes. Plants are shrubby, white woolly, 30-60 cm tall. The leaves are greyish green, aromatic, oblong, 7-8 cm long, on drying leaves turn silvery grey. They have soft velvet texture.

Useful Parts
Leaves.

Chemical Constituents
Protein, carbohydrates, minerals (calcium, phosphorus, iron, sodium, potassium) vit. A, vit. B_1, vit. B_2, volatile oil.

Processed Products
1. *Volatile Oil*: On steam-distillation, dry sage leaves yield an essential oil. The oil is produced commercially in Yugoslavia, Spain, Corfu, Syria and the USSR, and the samples from different sources show considerable variations in properties and constituents. The constituents present are a- pinene, cineole, linallacetate, thujone, borneol, bornyle acetate, farnesol and camphor. Linalyl acetate content of the oil is very low for the commercial exploitation of the oil for the ester. The quality of sage oil is determined by its thujone content. The higher the thujone content, the better is the oil. Leaves also contain tannin, fumaric, malic, and ursolic acids, a bitter principle, picrosalvin, saponin, pentoses, a wax and potassium nitrate.

2. *Oleoresin*: Sage oleoresin is dark green to brownish green. It is very viscous. It is usually extracted from Dalmatian sage. It is used as a natural antioxidant because of high phenol content.

Propertis and Uses

Flavour of sage varies from a mild, balsamic to a strong camphorous taste depending upon the variety. It provides cooling sensation in the mouth. It is used in pork sausage and baked loaf. The dried herb is grey, tinged with green; the aroma is strong, fragrant and spicy. The taste is fragrant, spicy, warm, astringent and a little bitter. It is used in every kitchen for flavouring meat and fish dishes and in making poultry stuffings. It is used in poultry dressing, sausage, liver sausage and hamburger seasoning in the western countries.

Sage is employed in the food industry as a standard spice in making stuffing for fowl, meats and sausage. It is one of the most important culinary herbs. Dried and powdered leaves are mixed with cooked vegetables. Thus are sprinkled on cheese dishes, cooked meats and other similar preparations. Fresh sage leaves are useful in salads and sandwitches. The young leaves are used as pickles and tea. Dried leaves are used in tooth and mouth-washes, gargles, poultices, tooth-powders, hair tonics and hair dressings. After steam-distillation, the residual plant material still contains constituents of considerable flavour value. Hot-water extracts of the material on concentration yield an oleoresin, which is used in conjunction with the oil for flavouring foods.

In Medicine

Sage is traditionally considered beneficial for throat and respiratory afflictions calming nerves and reducing fevers. It helps soothe digestion especially after heavy meal. It also strengthens gums and whitens teeth. Sage is a mild tonic, astringent and carminative. An infusion of the leaves is used as a gargle in the treatment of sore-throat. Hot infusion is diaphoretic. Extract of sage leaves is also reported to be antipyretic. A strong infusion of the herb is used to dry up the breast milk for weaning children. Sage oil finds use in perfumes as a deodorant, in insecticidal preparations, for the treatment of thrush and gingivitis and as carminative. The oil is used as a convulsant, and it resembles worm wood oil in action but is less active.

50. Savory
Satureia hortensis Linn.

Lamiaceae (Labiatae)

Savory is one of the most fragrant spices. It is indigenous to the mediterranean region. Savory is an erect pubescent annual herb, 25-35 cm in height with pinkish branches, found in Kashmir. Leaves are oblong-linear or lanceolate with deep-pitted glands on both sides. Flowers are in small axillary cymes. It grows in southern France, German, Spain and other parts of Europe; also in England, Canada and the USA. The dried leaves are of brown-green colour, and up to 10 mm in length. The odour is strong, warm and highly aromatic; the taste is somewhat sharp and camphoraceous. It is marketed both in whole and ground form.

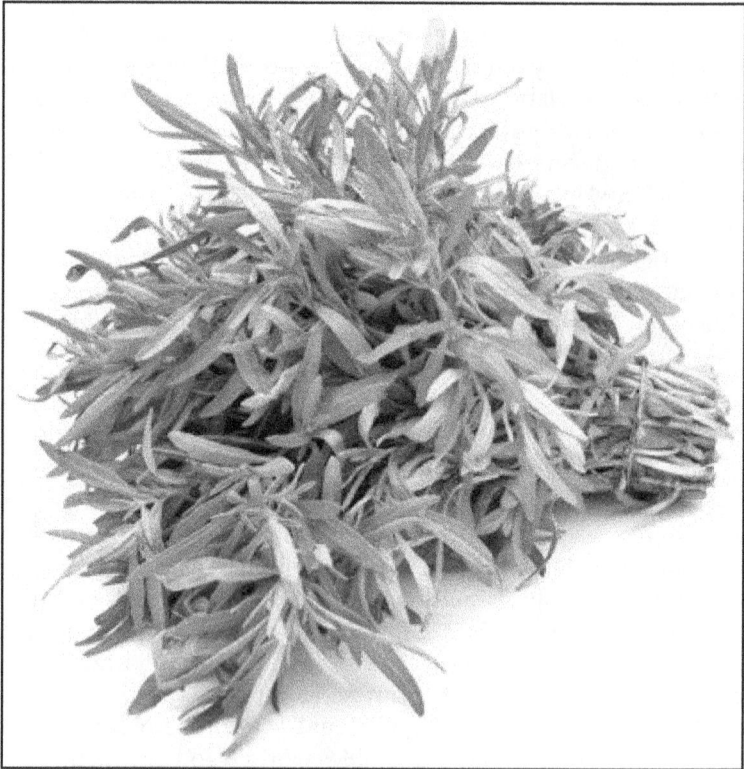

There are two species of *Satureia, viz., S. hortensis* Linn. And *S. montana* Linn., possessing a pronounced thyme-like odour and flavour. Both are employed as culinary herbs. The oil of the two plants is closely related in chemical composition. The dried leaves and flowering tops are derived from *S. hortensis* and constitute 'winter' savory of commerce. The best qualit savory comprises dried leaves. Only *S. hortensis* constitutes 'summer' savory.

Processing Technology

1. *Essential Oil*: Fresh flowering tops and leaves are used as both summer and winter savonries. The yield of volatile oil varies from 0.1 to 0.2 per cent from fresh leaves and flowering tops. The plant material is charged into a still. Steam generated in a separate steam generator is injected into the plant material and the oil is separated in a Florentune flask.

Useful Parts

Leaves, Flowering tops.

Chemical Constituents

1. *Dried Herb*: Protein, fat, carbohydrates, minerals (calcium, phosphorus, iron, sodium, potassium), vit. A, vit. B_2, vit. C, niacin.

2. *Fresh Herb*: Protein, fat, sugar, pentosans and minerals.

3. *Volatile Oil*: Savory contains 1 per cent volatile oil, chief constituent of which is carvacrol-a colourless to pale-yellow liquid with a pungent, thymol odour. The recovery of essential oil from the herb is higher during full flowering.

Propertis and Uses

Savory is used to flavour soups and sauces, egg, salad dishes, canned meats and poultry dressing. It is used in pork sausage and vegetables. Oil also find use in most of above food products.

In Medicine

The herb is reported to be carminative and stimulating. The anti-oxidant property of the herb reported in part is due to labiatic acid.

51. Shallot
Allium ascalonicum Linn.

Liliaceae

Shallot belongs to the onion or garlic family, the Liliaceae. It has similar properties to those of garlic. It is a native of Palestine and cultivated in USA, some European countries and in Indian gardens for its bulbs and green leaves. Clusters of greenish-white or red cloves occur at the base of its hollow cylindrical leaves. A variety of shallot, smaller and with a larger number of cloves, is exported to Sri Lanka and Malaysia from Madras.

Useful Parts

Leaves, Cloves.

Properties and Uses

Chiefly both leaves and cloves are used for flavouring curries. Most of the crop is produced for sale in the green state. Some dry bulbs are also used. Shallots are used for pickling.

In Medicine

It is used to cure ear-ache; a small piece being placed in the meatus. It is also fried in butter and preserved in honey as an aphorodisiac. The bulbs are ground and rubbed on the skins of feverish children in Ghana. Sometimes, they are mixed with palm-wine and large pepper and heated in the sun. The mixture is used to cure fever. They are also used as antidotes for snake bite and poisoning.

52. Spearmint
Mentha spicata Linn.

(Syn. *Mentha viridis* Linn.)
Lamiaceae (Labiatae)

Spearmint is a relative of mint and peppermint. It is a glabrous perennial 30-90 cm high, with creeping rhizomes, indigenous to the north of England. It is grown all-over the world. It is cultivated in

Indian gardens in plains. Leaves are smooth or nearly so, sessile, lanceolate to ovate, acute, coarsely dentate, and are smooth above and glandular below. Flowers are liliac, in loose, cylindrical, slender, interrupted spikes. The leaves have a characteristic aromatic odour and slightly pungent taste.

Useful Parts

Leaves, Flowering tops.

Chemical Constituents

Protein, fat, carbohydrates, minerals, calcium, phosphorus, iron, carotene (as vitamin A), nicotinic acid, riboflavin and thiamine, traces of copper, volatile oil (linalyl acitate, linalool).

Processed Products

1. *Volatile Oil*: The fresh flowering herb, on steam-distillation, yields 0.25-0.50 per cent of volatile oil, known as spearmint oil. It is a colourless, yellow or greenish-yellow liquid with the characteristic odour and taste of spearmint. The aroma improves on ageing.

   The main characteristic constituent of the oil is 1-carvone. It contains carvone, terpenes (chiefly a-limonene and dipentene),alcohol (as dihydrocarveol), and esters (as dihydrocarveol acetate). The essential oils obtained from Indian spearmint (*Mentha spicata*) subjected to different drying procedures were analysed. Forty-three flavour components were identified, of which eucarvone, cubenol and a-cadinol have been reported for the first time. The essential oil has a higher percentage of limonene (26.83 per cent) compared to Italian and Americn oils. Carvone to limonene ratios were 2.2:1.0 and 2.3:1.0 in the oil from fresh and shade-dried spearmint, respectively.

Properties and Uses
Green leaves of the plants are used for making chutney and for flavouring culinary preparations, vinegar, jellies and iced drinks. Spearmint oil is used for flavouring chewing-gums, tooth-pastes, confectionery and several pharmaceutical preparations.

In Medicine
The herb is stimulant, carminative and antispasmodic. A soothing tea is brewed from the leaves and an alcoholic beverage (mint julep) is prepared from them and it is then used as an antidote for poison. A sweetened infusion of the herb is given as a remedy for infantile troubles, vomiting in pregnancy and hysteria. The leaves benefit in fevers and bronchitis.

53. Star-Anise
Illicium verum Hook.

Magnoliaceae

Star-anise is the dried, star-shaped fruit. The tree is evergreen attaining a height of 8-15 m. Leaves are10-15 cm long and 2.5-5.0 cm broad and elliptic to oblanceolate. Fruits are star-shaped, reddish brown 6-8 follicles arranged in a whorl, radiating from the top of a slender short central stem. Each follicle is 12-17 mm long, boat-shaped, hard, and wrinkled and contains a seed. Seeds are brown, compressed-ovoid, smooth, shiny and brittle. The follicles have a pleasant anise-like odour and an aromatic, refreshing sweet, anise-like taste. The seeds have the same but to a much lesser extent. It is available whole, and is not usually ground.

The tree is indigenous to tropical and subtropical east Asia. It is widely cultivated in Kwangsi in south-east China and mainland

Tonkin in Indo-China. It is mostly imported from China and Indo-China. It is also produced recently in Arunachal Pradesh, India.

Useful Parts

Fruits.

Processed Products

Manufacturing Essential Oil

The fruits are distilled mostly in a fresh (green) condition. If so many fruits have accumulated at a distillation post, they may be kept for about 10 days, or even longer, provided they are spread out in a thin layer and frequently turned over, to prevent fermentation. Occasionally, the demand for dried fruits for export slackens and prices then fall to a very low level. Whenever this happens, the natives use their accumulated stokcs of dried fruits for distillation, hoping to obtain a better return from essential oil.

Production of the oil is entirely in native hands. The country-stills used for the purpose resemble those employed for distillation of cassia oil in China. They are primitive, although quite cleverly constructed. They hold up to 300 kg of fresh fruits per charge. In some regions, the natives break fruits by hand, prior to distillation. The fruits (in most still about 180 kg) are placed in the retort, together with sufficient water to cover the material. The heating by direct fire beneath the retort-vats is to be done slowly, to prevent boiling over of the charge and escape of vapours through the loose-fitting joints of the system. Steam and oil vapours pass from the retort through three small holes at the bottom of the vase-shaped still-head, which also serves as a sort of condenser. The still-head is covered with a flat bowl, through which cold water flows. Steam and oil vapours are condensed on the bottom of the flat bowl, the condensate drips into the lower part of the still-head, and from there, through a pipe, into the oil separator. Here, the oil and water separates; the distillation water flows back into the retort and is redistilled. Because of the small condenser surface, distillation in these stills must be carried out slowly. A charge of fresh fruit is required 48 hrs and a charge of dried fruit 60 hrs.

Native *vs* Modern Distillation of Oil

It is although primitive, production of star-anise oil in small native distillation posts offers however one great advantage, that of low-cost. The trees planted many years ago, grow semiwild near the villages and require no care. Harvesting carried out by village children, costs nothing. Any investment in simple still was amortized long ago. Distillation is a mere side occupation of villagers; it goes along, at a leisure place, with their principal task, that of growing rice. The male members of the family tend to their rice plots, casting an occasional glance at the flow of the distillate in their stills, and separating the oil; the women cook food near the still and kindle fire beneath it. In other words, the oil is produced at a practically no cost, and simply means additional income for the family.

Modern Distillery for Oil-Merits and Demerits

The erection of a modern distillery, on the other hand, would be relatively costly. Building and stills would have to be amortized, foreman and labourers to be hired at regular wages, fuel purchased, and taxes would have to be paid. True, in modern steam stills, the time of distillation could be shortened to perhaps 3 or 4 hrs, but the saving in cost of fuel would not compensate for other expenses involved. The only advantage would lie in the production of a very high grade of oil, free of all adulterants. Such an oil would consist almost entirely anethole and possess an exceptionally fine odour and flavour. The cheapest way to produce an oil of very high quality is simply to submit native-distilled oils to rectification, a practice which was followed by a large European dealers of star-anise oil in Langson, Tonkin, some years ago.

1. *Star-anise Oil*: The fruits whether fresh or dry are distilled using mostly native methods. It is colourless to pale-yellow. It has characteristic odour and aromatic taste more or less similar to true anise oil (from *Pimpinella anisum*). The oil contains anethole chiefly (85-90 per cent). Star-anise oil is official in the pharmacopias of many countries. It constitutes the bulk of oil of anise of commerce. The quality of star-anise oil, like that of aniseed oil, can be evaluated by its coagulating point.

2. *Leaf Oil*: The leaves of star-anise tree yield about 0.5 per cent essential oil on steam-distillation. It is inferior to that of fruit, both with respect to low congealing point and aroma.

3. *Fatty Oil*: Decorticated star-anise seeds contain fatty oil (55 per cent). The component fatty acids of the oil are myristic, stearic, oleic and lionoleic.

Properties and Uses

The fruits have an agreeable, aromatic, sweet taste and a pleasant odor, re-sembling anise. It is used as a condiment for flavouring curries, confectioneries and spirits, and for pickling. It is also used in perfumer. Star-anise oil is used as a flavouring agent in confectionery, candy, chewing gum, tobacco, animal feeds and liquors. It is also used in perfumery and soaps. Star-anise paste is used for flavouring foods and confectioner. The Chinese prepre a medicinal tea from leaves and also claim that one or two carpels of star-anise when added to chicken, which is to be roasted, improves its flavour tremendously.

In Medicine

It is stomachic and carminative and considered useful in flatulence, spasmodic affection of the intestinal canal and dysentery. It is used as an adjunct to cough, mixtures and as a corrective of taste. The fruit is chewed to sweeten breath and to help digestion. It is also used in production of absinthel (spirit).

The oil is stimulant, stomachic, carminative, mildly expectorant and diuretic. It relieves colic and is an ingredient of cough lozenges. The oil is administred in rheumatism and otalgia and as an antiseptic. It is useful against body lice and bed bugs and hence forms an ingredient of cattle sprays. It is used in vavus (honye-comb ringworm) and scabies. Star-anise paste is useful for colic, constipation and isomin.

54. Sweet Flag (Calamus)
Acorus calamus Linn.

Araceae

Sweet flag (calamus) is common in Kashmir Kumaon ranges of the Himalayan forests and Sri Lanka.

It is a semi-aquatic perennial herb. It has a creeping and much-branched aromatic cylindrical or slightly compressed rhizome about 19-25 mm in diameter and about 10 cm long. It is light brown or pinkish-brown, and is white and spongy within. It is the dried rhizome which constitutes calamus or sweet flag.

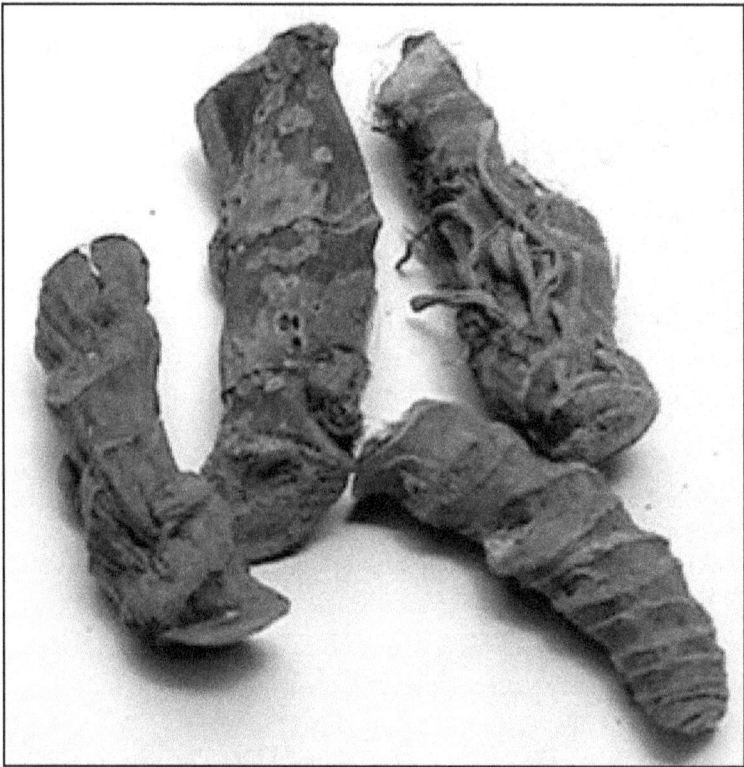

Useful Parts
Rhizomes.

Processed Products
1. *Volatile Oil*: The dry rhizomes on steam-distillation yield 1.5-3.5 per cent of a yellow aromatic volatile oil. It has a mellow odour resembling that of patchouli. The rhizomes are reduced to the desired size in a suitable disintegrator. It contain 'asarone'.

In Medicine
Its medicinal action is attributed to its essential oil content. The roots also contain glucoside called 'acorin'. The oil is reported to cure gastritis and is supposed to promote digestion also. The essential oil is used in stomach and skin diseases; as a vermifuge and antiseptic as well.

Infusion of the dried rhizome is aromatic, bitter tonic and carminative. It is emetic and antispasmodic. It is beneficial in cases of dyspepsia and chronic diarrhoea. It is also used in the treatment of insecticidal worms, the bites of poisonous reptiles, foul breath, throat infections, etc.

Propertis and Uses

The greatest use, at present, appears to be in the perfumer industry, as a component of the finest perfumes. It also enters into composition of face powders. The powdered rhizome is insecticidal and is used against bed-bugs, moths, lice, etc. It also forms the main ingredient of many insect powders, used to protect clothes, and are put in clothes and cupboards. The rhizome and its oil is used in the manufacture of liquor, essences and bitters.

55. Tamarind
Tamarindus indica Linn.

Caesalpiniaceae

Tamarind tree is one of the common and most important trees of India. Almost every part of the tree finds some use, but the most useful is its fruits. The ripe fruit (pod) of the tamarind tree is used as a condiment, or more precisely, as an 'acidulant'. India is perhaps one of the largest producers of tamarind in the world and it is the only country which produces a commercial crop of tamarind. It is chiefly grown in Madhya Pradesh, Andhra Pradesh, Tamil Nadu, Maharashtra and Karntaka.

It is a moderate-sized to large, evergreen tree, up to 24 m in height and found throughout the plains and sub-Himalayan tracts of India. Bark is brownish or dark grey. Pods are 7.5-20.0 cm long, about 2.5 cm broad and 1.0 cm thick, more or less constricted between seeds, slightly curved, brownish-colored, scurfy; seeds are 3-12, obovate-oblong, compressed, with a shallow, oblong pit on each side of the flat face and are of 1.5

cm 0.8 cm, smooth, dark brown and shining. The seeds are contained in loculi, enveloped by a tough, leathery endocarp. Outside the endocarp, is the light-brownish, red, sweetish, acidic, edible pulp traversed by a number of branched, ligneous strands. The outermost covering of the pod is fragile and is easily separable.

Useful Parts

Fruits.

Chemical Constituents

Pulp: Protein, fat, carbohydrates, minerals (Ca, P, Fe), carotene, riboflavin, niacin, vitamin C, tartaric acid, pectin, glucose, fructose.

Processed Products

Products like 'tamarind concentrate', tamarind powder', beverages', integrated process for the manufacture of pectin, tartarate and ethanol (alcohol) from tamarind pulp.

1. *Tamarind Concentrate*: Sauce manufacturers abroad use the tamarind extract for giving an exotic flavor to some sauces like Worcestershire sauce. The concentrate is found quite acceptable. It has a good flavour, is easy to dispense and reconstitutes well in hot water. The extraction of the pulp by boiling water dissolves the potassium bitartrate also which is not obtainable by cold-water extraction in the traditional way of using tamarind. While the pulp is subject to insect attack and proteolytic degradation during storage, the concentrate is clean and can be stored well for longer periods.

2. *Tamarind Pulp Powder*: Tamarind pulp is manually cleaned, de-seeded and fibres removed. The pulp is processed under standardized conditions of temperature, humidity and mill-settings in order to obtain an acceptable and hygienic product in powder form to facilitate its household use; possessing good reconstitutional characteristics. The product can be packed in the pouches or unit-packing Pulp is also used to obtain pectin, alcohol and tartarates.

3. *Tamarind Beverages*: Cooling beverages from tamarind pulp have been prepared and used particularly during summer months almost throughout India.

4. *Tamarind Seed Kernel Powder*: It is a valuable raw material for sizing of textiles. It is much cheaper than corn starch. It is also used as a good creaming agent for the concentration of rubber latex. It can also be used as a conditioner and stabilizer of soil for brick-making and as a binder for making sawdust briquettes. The powder is used as 'jellose' since the powder forms 'gel' with sugar and water at a specific pH, like pectin. But it is not a 'true pectin' since it does not contain glacturonic acid.

Properties and Uses

Ripe fruit or tamarind pulp has an important role in numerous culinary preparations in the country, notably sambhar, rasam, curries, chutneys, etc., which are quite popular throughout India.

In Medicine

It is beneficial in bilious vomiting, flatulence, indigestion, scurvy, cold, fevers, burns, etc. The ripe fruit is thought in Ayurveda as appetizing, laxative, healing; tonic to the heart, anthelmintic, cures 'vata' and 'kapha' heals wounds and fractures. The ripe fruit is refrigerant and digestive. It is useful in diseases caused by deranged bile, such as a burning of the body and intoxication from liquors or 'dhatura'. A gargle of tamarind water is recommended in sore-throat. Rasam and sambhar can do more than tickle your taste buds; they can protect you from fluorosis.

56. Tarragon
Artemisia dracunculus Linn.

Asteraceae (Compositae)

The dried leaves and flowering tops constitute what is known as 'Tarragon' (Estragon). It is so well-known to connoisseurs the world-over for its unusual flavour. It is also known as 'French Tarragon'. Its aroma is warm, aromatic and reminiscent of anise. It is a perennial herb found in western Tibet and in Lahaul, distributed in Afghanistan, Western Asia and South and Middle Russia, and cultivated in France and Spain temperate zones of the USA, as well as throughout the colder New England area.

Useful Parts

Leaves, Flowering tops.

Chemical Constituents

Methyl chavicol is the chief constituent of the volatile oil and phellandrene and ocimene are additional constituents. Dried leaves contain protein, fat, carbohydrates, minerals, calcium, phosphorus, iron, sodium, potassium, vit. A, vit. B. niacin, vit. C:

Processed Products

1. *Essential Oil*: Distillation of herb is carried out in direct steam-stills.

   Oil is a light-yellow to greenish liquid of peculiar, strongly aromatic odour, characteristic of plant and reminiscent of sweet basil and anise. Both yield and quality of essential oil are variable upon the basic quality of the raw material used, method and duration of distillation.

Properties and Uses

It is used to flavour vinegar ('terragon vinegar') pickles and prepared mustard. It is also used to flavour soups, salads, meat dishes, certain cheeses, canned soups, table sauces, salad dressings and liqueur. It is also employed in perfumes.

In Medicine

The aromatic leaves are credited with aperient, stomachic, stimulant and febrifuge properties.

57. Thyme
Thymus vulgaris Linn.

Syn. *Thymus serpyllum* Linn. (Wild Thyme)
Lamiaceae (Labiatae)

Thyme is one of the best known European culinary herb. The dried leaves and flowering tops of *T. vulgaris* Linn. are called 'thyme' and those of *T. serpyllum* Linn. Are known as 'Wild thyme'. The latter is found in the western temperate Himalayas, from Kashmir to Kumaon hills at an altitude of 1,525 m above sea level. The former is grown in Europe, Australia and North Asia. It is now cultivated in France, Germany, Spain, Italy and other parts of Europe and also in England, North Africa, Canada and the USA. It is a common garden-plant. It survives for many years. The dried leaves are curled brownish-green upto 6-7 mm, 2-3 mm broad and marketed in whole or ground form. The flavour is aromatic, warm and pungent.

The latter is a low, hard perennial undershrub, 20-40 cm high, successfully introduced in the Nilgiris at higher elevations. It is used both for seasoning and for its volatile oil. Leaves are oblong-lanceolate, 10 mm x 3 mm, with orange-brown, glandular dots. The flowers are coriaceous, small, lavender, purplish or bluish to nearly white in verticillasters, nutlets brown.

Useful Parts
Leaves, Flowering tops.

Chemical Constituents
Protein, fat, carbohydrates, minerals (calcium, phosphorus, iron, sodium, potassium) Vit. A, Vit B_1, Vit B_2, Vit C, niacin.

Processed Products
1. *Volatile Oil*: Essential oil and oleoresin both are prepared from the flowering tops. The volatile oil is the most important extractive. The Spanish red thyme is the prime

source of the oil. It is marketed as a 'Red Thyme Oil'. There is also'white thyme oil' but this is a redistillation product of the red oil rather than the product of another plant. It tends to be smoother in flavor than the red oil and pleasingly aromatic. Crushed thyme, on steam-distillation, yields about 2.5 per cent volatile oil which is a colourless to yellowish-reddish liquid with a pleasant odour. The chief constituent of the oil is thymol (about 45 per cent). In the pure state, thymol consists of colourless, translucent crystals possessing an aromatic thyme-like odour and a pungent taste.

2. *Oleoresin of Thyme*: It contains both volatile and non-volatile components. It is a very viscous material which is normally standardized with a food-grade solvent before use. The oleoresin is often added to soluble carriers, either liquid or dry, for seasoning applications.

Propertis and Uses

The herb is used to season tomato soups, 'clam chowder' and juice, fish and meat dishes. It is also used in liver sausage, pork sausage, head cheese, cottage and cream cheese and bockwurst.

In Medicine

The oil of thyme is used in the treatment of bronchitis and whooping cough. The oil of wild thyme is used in toothache. In Punjab, the herb of wild thyme is given in week vision, complaints of stomach and liver, supression of urine and menstruation. In Europe, it is considered anti-spasmodic and carminative. An infusion of herb is given in convulsive coughs, whooping coughs, catarrh and sore throat. It is good for nervous or hysterical headaches and for flatulence. The herb is emmenagogue, alexiteric, anthelmintic; good in liver complaints, pain in spleen, liver or chest. It is useful in asthma and bronchitis in thins phlegm and blood. The leaves are laxative, stomachic and tonic. It is good for the kidney and eye, and are blood-purifier. It soothes the throat and cures the cough. It strengthens the lungs, promotes sleep and rids nightmares; improves digestion, cures a hangover, helps the dull sightedness; warms the heart; soothes the liver and takes away hardness of the spleen. It is excellent for shortness of breath and provokes the terms. It takes away hot swelling; eases the gout; cures pains in the loins and hips; and purges the phlegm.

58. True Cardamom (Small Cardamom)
Elettaria cardamomum (Linn.) Maton

Zingiberaceae

The true cardamom of commerce is 'Lesser cardamom', 'Green cardamom', 'Malabar cardamom', 'Small cardamon' or *'Chhoti elaichi'*. It is popularly known as the 'Queen of Spices'. It is one of the most valued spices of the world. Among spices of India, it occupies an eminent position as the second largest foreign-exchange earner. Greeks, Romans and Chinese used over 2000 year ago in food,

medicines and perfumes. The Vikings bought cardamom from traders who obtained it through Constantinople and introduced in Scandinavia. Cleopatra filled her chambers with the sweet smell of cardamom smoke before Mark Anthony's visit to Egypt.

True cardamoms are normally marketed in four different forms on the basis of their geographical origin and on their physical forms. Some users express strong preference for particular types in certain applications. For example: (i) *Whole green cardamoms*: These are mature fruits picked while still green, their colour having been preserved by special treatment (steeping in dilute solution of sodium carbonate for a specific period). (ii) *Whole bleached cardamoms*: Mature fruits picked after the green colour has faded and which have been bleached with sulphur dioxide to produce a uniform white appearance. (iii) *Whole, straw coloured cardamoms*: Mature fruits which have been simply dried in the sun. (iv) *Decorticated seeds:* This tends to be a product of expediency where due to poor processing capsules have been blemished or split.

Whole cardamoms are categorized as Malabar and Mysore type or race on the basis of their appearance and chemical characteristics:

1. *Malabar type*: This race has the most pleasant and mellow aroma and flavour character. It is, but, not so well fit for the preparation of green cardamom, for when the seeds have developed their full flavour character, the capsules being to turn pale yellow. The green colour cannot be regained during curing. The cured fruits are generally rounded and are about 18.5mm in length.

2. *Mysore type*: This race has a somewhat harsher aroma and flavour. It is the most widely grown since the seeds reach flavour maturation when the capsules are still green, which facilitates preparation of green spice. The cured fruits are 3-cornered and ribbed and tend to be slightly longer than Malabar type, being about 21 mm in length. Both races are indigenous to India but both types have been introduced into other countries. The higher cineol content of Malbar cardamom oil makes it more suitable for certain uses. Conversely, the lower cineol content of the Malbar cardamom oil makes it more suitable for other purposes.

Substitute Spices (False Cardamoms)

There are two species which usually found in the western trade as low grade substitutes for 'true cardamom' *viz.* (1) *A. subulatum* Roxb., the Nepal cardamom, grown in Nepal, Sikkim, West Bengal and Bhutan and (2) *A. korarima* Pereira, the Kararima cardamom from Ethiopia. The wild cardamom from Sri Lanka *viz., E. cardamomum* var. 'Major Thwaites' enters trade to some extent as an adulteration in consignments of decorticated seeds of true cardamom or true cardamom oil.

Useful Parts

Fruits and seeds.

Chemical Constituents

Volatile oil, essential oil (cineole, α-terpinol acetate, linalyl acetate, sabinene, limonene, linalool) and minerals (Ca, Mg, K, Mn), Vitamins A, B_1, B_2 and C.

Properties and Uses

True cardamom is used directly as a flavouring material generally in three forms: i)whole, ii)decorticated seeds and iii)ground. The spice is also processed on an industrial scale to prepare distilled essential oil and the solvent-extracted oleoresin. True cardamom is marketed mostly in whole form. Trade in decorticated seeds is much smaller and it is negligible in the ground spices.

As Masticatory and Food Flavourant

The principal use of true cardamoms worldover is for the sake of domestic culinary purposes in whole or ground form. The spice plays an important role in a variety of spiced rice, vegetable and meat dishes. Indian cardamom adds a lingering sparkle to every kind of cooking or dishes in Asia, both traditional and modern. In Arab countries of the Middle East, the spice has been traditionally used for flavouring coffee, and in Scandinavia, for flavouring range of baked goods, including cakes, buns, pastries and bread. In other European countries and in North America, the spice is used mainly in ground form by food industries as an ingredient in cury powder, some sausage products, soups, canned fish and to a small extent in flavoring of tobacco. Cardamom is used in cola and instant gahwa, carbonated gahwa, biscuits, Danish pastries, toffees, chewing gum, encapsulated cardamom oil and various breakfast foods. The Arabs employ it in coffee; the Americans do so in baked foods; the Russians in pastries, cakes and confectionary; the Japanese in curry, ham and sausage; the Germans in curry, powders, sausage and processed meats; the Scandinavians in coffee and cakes. In India, seed powder is added during cooking or sprinkled on savory dishes to give a sweet aroma. Green cardamom is an essential ingredient in Indian sweets, puddings, ice-creams, etc. It is added in hot tea and in many as blend.

Cardamom Oil

The essential oil finds place in flavouring of processed food. It is also used in certain liquid products such as cordials, bitters and liqueurs and sometimes in perfumery.

Cardamom Oleoresin

It is used in flavoring of processed foods but less extensively. Both the oil and oleoresin tend to develop 'off flavours' when exposed to air for prolonged period. Their usage is usually restricted to meat products of intended short shelf-life such as sausages.

In Medicine

Tinctures of cardamom are used chiefly in medicines for windiness or stomachic. Powdered cardamom seeds along with ground ginger, cloves and caraway are beneficial in combating digestive ailments. It is used as a powerful aromatic stimulant, carminative, stomachic and diuretic. It checks nausea and vomiting. It is reported to be a cardiac stimulant. A good nasal application is prepared by using extract of cardamom, neem and myrobalan along with animal fat and camphor. Cardamom seeds are chewed to prevent bad smell in mouth, indigestion, nausea and vomiting due to morning sickness, excessive watering in mouth (pyrosis), etc.

Gargling with the infusion of cardamom and cinnamon is remedial against pharyngitis, sore throat, hoarseness during infective stage of flu. Its daily gargle protects one from flu.

Powdered seeds of cardamom boiled with tea-water imparts a very pleasant aroma to tea. This is administered for scanty urination, diarrhea, dysentery, palpitation of heart, exhaustion due to over-work, depression, etc.

Eating a cardamom once daily with a tablespoon of honey betters eye-sight, strengthens nervous system and render healthy. It is employed in a variety of ailments such as smell of mouth, gonorrhoea, cystitis, nephritis, burning micturition, scanty urination, impotency, premature ejaculation, depression, hiccup pharygitis, sore throat, relaxed uvula, headache, kidney stones, etc.

Processed Products

1. *Essential Oil*: Cardamom oil is produced commercially by steam distillation of crushed fruits. The yield and organoleptic properties of the essential oil so obtained are dependent upon the type of the spice used. The best oil yields are obtained with Alleppey Greens, but the manufacturer's choice of the type of material takes into account the economics of oil recovery. Very little has been published on the technique of cardamom oil distillation, except that the necessity of complete distillation to obtain full flavour character of oil has been pointed out. It has been also shown that at least four hours' distillation is required to produce the full ester content of the oil.

2. *Oleoresin*: Cardamom oleoresin is produced on a relatively small scale. Solvent extraction yields are of the order of 10 per cent and the content of volatile oil in the oleoresin is dependent upon the raw material and solvent used. Cardamom oleoresin used for food flavouring is normally dispesed on salt, flour, rusk or dextrose.

3. *Decorticated Seeds and Seed Powder*: Decorticated cardamom seed generally have lower price than whole cardamoms. This is so because rapid loss volatile oil during storage and transportation. Seeds are also classified by weight into 'prime' and 'light' seeds. The former fetches a better price. Seed powder is sold on a limited scale.

59. Turmeric
Curcuma longa Linn.

(syn. *C. domestica* Val.)
Zingiberaceae

Turmeric is a native of Southern or South – Eastern Asia. It is mentioned in many Sanskrit treatise. It seems to have reached China before the 4th century. It spread throughout the East Indies and eastwards across the pacific by polynesians as far as Hawaii and Easter, Island. It was known to Dioscorides (77 A.D.) and Marco Polo (1280).

The spice 'turmeric' or *'haldi'* constitutes boiled, dried, cleaned and polished rhizomes of *Curcuma longa* Linn. The plant, is now a commercial crop of tropics. Turmeric is one of the most important and ancient spices of India and a traditional item of export. It is cultivated extensively in India, Sri Lanka, parts of China, Indo-China and Pakistan. India is the largest producer and also exporter of turmeric in the world. In India, the turmeric is usually cultivated in Andhra Pradesh, Maharashtra, Odisha, Tamil Nadu, Karnataka and Kerala. The cultivated varieties go by the name of localities *e.g.* Duggirala, Tekkurpet, Amalapuraii, Alleppey, Lakadong, etc.

Useful Parts

Rhizomes.

Chemical Constituents

Curcumin, Volatile oil, Essential oil (turmerones, zingiberene, ketone) oleoresin, minerals (Ca, P, Fe), carotene, thiamine, niacin.

Processed Products

1. *Turmeric Powder*: Turmeric is in the form of hard fingers in trade. It is powdered in two stages to get a fine powder of about 60 mesh. Turmeric is prized for its aroma and colour. The colouring matter is stable to heat. It, therefore, does not pose much problem during grinding ever though heat is generated. It is an important constituent in curry powder formulations, pickles and soups.

   Its powder is fumigated to prevent insect infestation. It is packed in laminate pouches of cello/poly or metallized polyester/poly or paper/foil/poly to keep well for a year. For short-term storage for about 90 days, 200-gauge polyethylene or high density polyethylene be used.

   The powder is mostly consumed domestically in culinary preparations. It is also used in the manufacture of curry powders and various spice mixtures.

2. *Volatile Oil*: The volatile oil is derived by steam distillation of crushed turmeric rhizome. It is an orange-yellow liquid, occasionally slightl fluorescent with an odour. The dried rhizomes yield 5-6 per cent of oil and fresh ones give 0.24 per cent essential oil. About 58 per cent of the oil is composed of turmerones (sesquiterpene ketones) and 9 per cent tertiary alcohols.

3. *Oleoresin*: A standardized technique for the manufacture of oleoresin from ground turmeric by solvent extraction is followed by vaccum concentration. The semiliquid viscous stuff contains volatile aromatic principles and non-volatile acrid fraction as well. This covers the overall aroma and flavour in a concentrated form without starch and fibrous contents. It is in great demand by the food and pharmaceutical industries abroad. Volatile oil content of turmeric ranges from 2.5 to 7.2 per cent and their curcumin content from 1.8 to 5.4 per cent. The bright-yellow colour of turmeric found by thin-layer chromatography was due to curcumin and to two other related pigments with the same absorption maximum (425 nm) in the visible range.

Acetone, alcohol and ethylene dichloride are found suitable for extracting oleoresin from turmeric. Oleoresin is a highly viscous orange-brown product containing 30-35 per cent curcumin, 15-20 per cent volatile oil and has a characteristic turmeric aroma. Turmeric oleoresin finds application in meat and fish products, pickles, dairy products and confectionery.

Properties and Uses

The largest amount of turmeric is utilized in most of the Asiatic countries as a food adjunct in vegetable, meat and fish preparations. It is traditionally used to flavour and at the same time to colour butter cheese, margarine, pickles, mustard and other foodstuffs. It is also used to colour liquor, fruit drinks, cakes and table jellies. It is one of the principle ingredients of the curry powder. A pinch of turmeric powder is often added to most of our savouries to impart simultaneously an agreeable flavour and colour and to improve keeping quality. It is still used for dyeing cotton. The dye is also employed as colouring material in pharmacy, confectionery, rice milling and food industries. Turmeric is converted as 'kumkum' for 'tilak'. Turmeric and its preparations like 'kumkum' and 'parani' serve Hindu women as an inexpensive and indigenous beauty aid. Smearing turmeric paste on face and limbs during a bath is found to clear the skin and beautiful face. It is useful in the manufacture of paints and varnishes.

In Medicine

Turmeric finds an important place, as an ingredient, in the preparation of medicinal oils, ointments and poultice. In the Indian systems of medicine – Ayurvedic, it is considered stomachic, carminative, tonic, blood purifier, vermicide and an antiseptic. Its antiseptic and healing properties are said to be both a preventive cure for pimples. It is also useful to discourage unwanted hair on feminine skin. It is also prescribed as an antiperiodic alternative. It relieves sore throat and common cold, if taken internally in the form of warm milk or inhalved from boiling water or as a smoke.

The juice of raw rhizomes is used as an anti-parasitic for many skin affections. In small-pox, it is applied as a paste with gingelly (sesame) oil and neem leaves. Burnt turmeric used as tooth powder relives dental troubles. Turmeric is useful in treating gall stones and gall complaints. The essential oil of turmeric is antiseptic.

Turmeric is all-round wonder spice. It help detoxify the liver, balance cholesterol levels, fight allergies, stimulate digestion and boost immunity. It is also reported beneficial in the treatments of malaria, cough, phlegm and jaundice. It benefits in a wide range of ailments *e.g.* bronchial asthma, anaemia, arthritis, measles, cold and cough, ringworm, scabies, foul breath and dental caries, etc.

60. Vanilla
Vanilla fragrans (Salisbury) Yames

Syn. Vanilla planifolia Andrews
Orchidaceae

Vanilla was used in pre-Columbian times by the Aztecs of America as a flavouring for chocolate as a source of perfume and as herbal tonic. It was also known to have been used as a medium of exchange for paying tribute to the ruler. Bernal Diaz (the leader of the Spanish conquistador) was perhaps the first European to observe vanilla spice in making of chocolate drink, which was used by emperor Montezumma. The vanilla pods "sticks of commerce" are

the cured fruits of the climbing orchid *Vanilla fragrans*. It is a native of Atlantic Coast from Mexico to Brazil. Vanilla cultivation had spread to other countries after the discovery of America. The important vanilla-producing countries are: Madagascar, Mexico, Tahiti, Malagasy Republic, Comoro, Reunion, Indonesia, etc. Vanilla constitutes one of the most important flavours used in food, perfumery and pharmaceutical industries.

Vanilla is known commercially under various names such as 'Burbon vanilla', Mexican vanilla', 'Indonesian vanilla' and

'Seychelles vanilla'. Four commercial forms are well established in the international trade:(i) Vanilla pods, consisting of whole pods which may be split. (ii) Cut vanilla, consisting of parts of pods, split or not and deliberately cut or broken. (iii) Vanilla in bulk, consisting of vanilla in pods and cut vanilla. (iv) Vanilla powder, obtained by grinding vanilla pods, without additives and after drying.

Useful Parts

Fruits.

Chemical Constituents

Vanilla beans: protein, fatty oil, volatile oil, nitrogen-free extract, carbohydrates, vanillin, resins, minerals (calcium, potassium, sodium, phosphorus, iron).

Processed Products

1. *Vanilla Extracts/Essences*: Vanilla extracts are in great demand in America, while cured vanilla beans are popular in European countries. The vanilla flavour can be extracted with alcohol. The colour of the extract depends on the strength of the alcohol used, the duration of extracting and the presence of the glycerin. Dark-coloured extract is obtained from dry beans and the presence of glycerin deepens the colour of the extract. Vanilla extract is stored in stainless, aluminium or glass containers. Ageing for 25-30 days improves the aroma due to formation of esters from acids in the presence of 42-45 per cent alcohol. Wooden container should be avoided completely as it adversely affects its flavour because of the alcoholic extractions from the wood itself.

2. *Vanilla Sugar*: The vanilla extract is mixed with sugar and it is made into a powder called 'powdered vanilla' or 'vanilla sugar'.

3. *Vanilla Powder, 'Vanilla Absolute' and Vanilla Tincture*: They are used for perfumery. These are produced and marketed in the USA and few other countries.

Propertis and Uses

Vanilla, today, constitutes the world's most popular flavouring for numerous sweetened food. 'Vanilla Sugar' is used in the manufacture of chocolates. 'Vanilla flavouring' is used in liquor, in cheap brandy and in whisky. In the USA, Vanilla flavour is widely used as a flavouring for ice-creams, soft drinks, chocolate, confectionery, candy, tobacco, baked foods, puddings, cakes, cookies and liquors. It is also used as fragrantly tenaceous ingredient of perfumery. About 50 per cent of vanilla is used in ice-cream industry alone.

Bibliography

Abd-El-dHamide, M.F., Atalah, R.K. and Moussa, Z.A., 1984. Chemical studies on Egyptian fenugreek, *Trigonella foenum-graecum*. *Annals Agric. Sci. in Shams Univ. Chashenut*, 29(1): 43–60.

Abraham, K.O., Shankaranarayana, M.L., Raghavan, B. and Natarajan, C.P., 1979. Asafoetida. IV. Studies on volatile oil. *Indian Fd. Packer*, 33(1): 29–32.

Abraham, K.O., Shankaranarayana, M.L., Raghavan, B. and Natarajan, C.P., 1982. Odorous compounds of Asafoetida. VIII. Isolation and identification. *Indian Fd. Packer*, 36(5): 65–76.

Akhtar, M.A., Nasir, K. Ashraf, M., Moinnuddin, M. and Khan, M.R., 1982. Physio-chemical study of essential oil from parsley seeds. *J. Nat. Sci. Math*, 22(2): 81–88.

Anjoe, K., Lonnerdal, B., Uppstrom, B. and Aman, P., 1997. Composition of seeds from *Brassica* cultivars. *Swedish J. Agric.*, 7(4): 169–178.

Anonymous, 1950. Ajowan. In: *Wealth of India–Raw Materials Vol. 10*. Publication Department, CSIR, New Delhi, India., pp. 267–271.

Anonymous, 1950. *Crocus sativus*. In: *Wealth of India–Raw Materials Vol. 2*. Publication Department, CSIR, New Delhi, India, pp. 370–372.

Anonymous, 1950. Leek. In: *Wealth of India–Raw Materials Vol. 1*, 2nd edn. Publication Department, CSIR, New Delhi, India.

Anonymous, 1950. *Wealth of India–Raw Materials Vol. 2*. Publication Department, CSIR, New Delhi, India.

Anonymous, 1962. Curry leaf. In: *Wealth of India–Raw Materials Vol. 6*. Publication Department, CSIR, New Delhi, India, pp. 481.

Anonymous, 1962. Majoram. In: *Wealth of India–Raw Materials Vol. 6*. Publication Department, CSIR, New Delhi, India, pp. 226–227.

Anonymous, 1963. *Vanilla*. Indian Council of Agric. Res., New Delhi, Bull No. 23, India.

Anonymous, 1966. *Wealth of India–Raw Materials Vol. 7*. Publication Department, CSIR, New Delhi, India.

Anonymous, 1970. *Wealth of India–Raw Materials Vol. 1 (A–B)*. Publication Department, CSIR, New Delhi, India.

Anonymous, 1970. *Wealth of India–Raw Materials Vol. 10*, Publication Department, CSIR, New Delhi, India, pp. 93–99.

Anonymous, 1970. *Wealth of India–Raw Materials Vol. 5*. Publication Department, CSIR, New Delhi, India.

Anonymous, 1984. Mustard powder from mustard seeds by CFTRI improved process. In: NRDC process No. 1019.09.84. National Research and Development Corporation of India, New Delhi, India.

Anonymous, 1985. *Acorus calamus* Linn. In: *Wealth of India–Raw Materials Vol. 1(A)*. Publication Department, CSIR, New Delhi, India, pp. 63–65.

Anonymous, 2011. *The Complete Book on Spices and Condiments with Cultivation, Processing and Uses*. NIIR Board of Consultants and Engineers, Asia Pacific Business Press, Delhi, India.

Arana, F.E., 1946. Vanilla curing and its chemistry USDA Fed. Expt. Station, Maugulz. Puerto Rico. Bull. No. 42, Washington DC USDA.

Ashraf, M., Sandra, P.J., Saeed, T. and Bhatt, M.K., 1979. Studies on the essential oils of the Pakistani species of the family Umbellifrae–II. *Petroselinum crispum* Millet (Parsley) seed oil. *Pak J. Sci. Ind. Res.*, 22(5): 262–264.

Ashurst, P.R., Lee, D.A. and Lewis, O.M., 1972. The development of oils of pimento. *J. Sci. Res. Council,* Jamaica, 3(1): 50.

Aykroyd, W.R., Wardhan, V.N. and Ranganathan, S., 1991. *The Nutritive Value of Indian Foods and the Planning of Satisfactory Diets.* Manager of Publications, Govt. of India, Delhi, India.

Bakhru, H.K., 2001. *Indian Spices and Condiments: As Natural Healers.* Jaico Publishing House, Mumbai, India.

Balbaa, S.L., Hilal, S.H. and Haggage, M.Y., 1975. A study of the fixed oils of the fruits of *Carum copticum* Benth. and Hook, *Apium graveolens* L. and *Petroselinum sativum* Hoffm. growing in Egypt. *Egypt J. Pharm. Sci.,* 16 (3): 383–390.

Balbi, G., 1960. Poppy seed oil in the manufacture of pectin and varnishes. *Olearia,* 14: 97–104.

Baldry, J., Dougan, J., Mathews, W.S., Nabney, J., Pickenns, G.R. and Robinson, F.V., 1976. Composition flavour of nutmeg oils. *Flavour,* 7: 28–30.

Baldry, J., Mathews, W.S., Robinson, F.V. and Nabney, J., 1974. Chemical composition and flavour of nutmegs of different geographical regions IV. *International Congress of Food Science and Technology,* 19: 38–40.

Bali, A.S. and Sagwal, S.S., 1987. Saffron: A cash crop of Kashmir. *Agriculture Situation in India,* 41(2): 965–968.

Baser, K.H.C., Ozek, T., Tumen, G. and Sezik, E., 1993. Composition of the essential oils of Turkish *Origanum* species with commercial importance. *Journal of Essential oil Research,* 5(6): 619.

Basker, D. and Negbi, M., 1983. Uses of saffron. *Econ. Bot.,* 37: 228–236.

Basls, R.K., 1977. Essential oil of *M. piperita* raised in the Kumaon region. *Natl. Appl. Sci. Bull.,* 29(2): 75–76.

Basls, R.K., Gupta, R. and Baslas, K.K., 1971. Chemistry constituents of essential oils from the genus *Anethum* (Umbelliferae), Oil of seed of *Anethum graveolens* part I. *Flavour Ind.,* 2: 241–245.

Bhartiya, S.P., 1967. Kal Zira is a good bet. *Indian Fmg.,* 17 (1): 30–32.

Bhown, A.S., Shah, D.K. and Nath, S., 1965. Studies on poppy seed (*Papaver sommiforum*) part I. Free amino acid. *Naturwissenschaften,* 52(18): 516–517.

Boatto, G., Pintore, G., Palomba, M., Simone, F. De, Ramundo, E. and Iodice, C., 1994. Composition and antimicrobial activity of *Rosmarinus officinalis* essential oils. *Fitoterapia*, 65(3): 279–280.

Brown *et al.*, 1955–1956. Cinnamon and Cassia: Soun production and trade. Col. P1. *Anim. Prod.*, 5: 257–80, 6: 96–116.

Chakravarti, K.K. and Bhattacharya, S.C., 1934. Chemical examination of Indian spearmint oil. Part I. *Perfume Essential Oil Rec.*, 45: 217–218.

Cheema, G.S., Bhat, S.S. and Naik, K.C., 1954. *Commercial Fruits of India*. Macmillan and Co. Ltd., London, pp. 270–284.

Choudhari, B., 1983. *Vegetable,* 7th Repr. National Book Trust, New Delhi, India.

Choudhury, B., 1983. Cele. In: *Vegetables*. National Book Trust, New Delhi, India, pp. 180–182.

Choudhury, B., 1983. Leek. In: *Vegetables*. National Book Trust, New Delhi, India, pp. 105–106.

Cui, W., Eskin, N.A.M. and Bihaderis, C.G., 1993. Chemical and physical properties of yellow mustard mucilase. *Food Chemistry*, 46(2): 169–177.

Dahanukar, S.A., Karandikar S. and Desai, M., 1984. Efficacy of *P. longum* in childhood asthma. *Indian Drugs*, 21(9): 384–388.

El-Khrisy, E.A.M., Mahmoud, A.M. and Abu-Mustafa, E.A., 1980. Chemical constituents of *Foemiculum vulage* fruits. *Fitoterapia* SI, (5): 272–275.

Embong, M.B. and Hudziyev, D., 1977. Essential oil from species grown in Alberta Dill seed oil (*Anethum graveolens* L.). *Can Inst. Fd. Sci. Technol. J.*, 10(3): 208–214.

Embong, M.B., Hadziyev, D. and Molnar, S., 1977. Essential oils from spices grown in *Alberca carawa* Essential oils (*Carum Carvi*). *Can. J. Pl. Sci.*, (2): 543–549.

Fleisher A., 1980. Essential oil from two varieties of *Ocimum basilicum* L. grown in Israel. *J. Sci. Fd. Agric.*, 32(4): 119–122.

Garg, S.K., Gupta, S.R. and Sharma, N.D., 1979. Coumarins from *Apjum graveolens* seeds. *Phytochemistry*, 18(9): 1580–1581.

Garg, S.K., Gupta, S.R. and Sharma, N.D., 1979. Minor phenolies of *Apium graveolens* seeds. *Phytochemistry*, 18(2): 353.

Gopalkrishnan, M and Varkey, George, 1980. Cinnamon barks from Mainpur region of India. *Indian Cocoa, Arecanut and Spices J.*, 3(3): 65.

Gopalkrishnan, M., 1979. Chemical aspects of clarification of Cinnamon oil. *Indian Perfumer*, 23: 2.

Gopalkrishnan, M., 1992. Chemical composition of nutmeg and mace. *J. Spices and Aromatic Crops*, 1(1): 49–54.

Gopalkrishnan, M., Padmakumri, K.P., Narayana, C.S. and Mathew, A.G., (1984. Lipid components of clove (*Syzygium aromaticum*) bud oil extracted using carbon diozide. *Journal of the Science of Food and Agriculture*, 5011: 11–119.

Gopalkrishnan, M., Padmakumri, K.P., Narayana, C.S. and Mathew, A.G., 1984. Lipid components of clove oil. *J. Fd. Sci. Technol.*, India, 2(1): 52.

Gowda, P.H.R., Shiva Shankar, K.T. and Gowda, I.V.N., 1990. Uses and cultivation of curry leaf. *Spice India*, 3(2): 7.

Guenther, E., 1972. *The Essential Oils* (5th Repr). D. Van Nostrand, New York, U.S.A.

Gulati, B.C., 1982. Essential oils of *Cinnamonum* species. In: *Cultivation and Utilisation of Aromatic Plants*, (Eds.) C.K. Atal and B.M. Kapur. Publication and Information Directorate, CSIR, New Delhi, India, pp. 607–629.

Gupta, G.K., Dhar, K.I. and Atal, C.K., 1977. Chemical constitution of *Coriandrum sativum* Linn. *Indian Perfumer*, 21: 86–90.

Halim, A.G. and Ross, S.A., 1977. Flaveroids of cumin. *Egypt J. Pharm. Sci.*, 18(3): 245–252.

Handa, K.I. *et al.*, 1963. Essential oil of *P. longum* properties of the components and isolation of two monocyclic sesquiterpenes. *Perfume Kosmetick*, 44(9): 233–236.

Handon, W.I. and Hocking, G.M., 1957. Sage and garden sage. *Econ. Bot.*, 11(1): 64–67.

Hiremath, S.M., Madalagiri, B.B. and Basarkar, P.W., 1996. Essential oil constituents for wild curry leaf. *Indian Perfumer*, 40(4): 110–112.

Huopalathi, R., 1986. The contents and composition of aroma compounds in three different cultivers of Dill. *Z. Lebensmittelunters Forsch*, 18(2): 57–64.

Hussain, A. *et al.*, 1998. Clary sage oil. In: *Major Essential Oil Bearing Plants of India*, CIMAP, Lucknow, pp. 58–61.

Ina, K., Sano, A., Nobukni, M. and Krishna, T., 1981. Studies on the volatile compounds of Lorse-radish. *J. Jap Soc. Fd. Sci. and Tech.*, 28(7): 365–370.

Jain, M.L. and Jain, S.R., 1972. Therapeutic utility of *O. basilicum* var. *album. Planta Med.*, 22(1): 66.

Jones, H.A. and Mann, I.K., 1963. *Onions and their Allies*. World Crop Books, Intersci Pub. Inc. New York, U.S.A.

Joseph J., 1980. The nugtmeg: Its botany, agronomy, production, composition and uses. *J. Plntn. Crops*, 8(2): 61–72.

Kaaner, R.E., 1985. Antiozidnts in clove. *J. Am. Oil Chem. Soc.*, 62(1): 111–113.

Karawya, M.S., Hinawy, M.S. and El-Hawary, S.S., 1977. Egyptian essential oil of *Mentha piperida* of different stages of plant maturity. *Egypt J. Pharm. Sci.*, 18(4): 387–394.

Karim, A. and Bhutty, M.K., 1976. Studies on the essential oils of the Pakistani species of the family Umbelliferae. IV *Apium graveolens* Linn. (Celer ajmodh) seed oil. *Pakist J. Scient. Ind. Res.*, 119(56): 243–246.

Karim, A. Shraf, M. Pervez, M. and Bhutty, M.K., 1977. Studies on the essential oils of the Pakistani species of the family Umbelliferae–VIII. *Carum carvi* Linn. oil of the immature plant. *Pakistan J. Scient. Ind. Res.*, p. 20.

Katyal, S.I., 1977. *Vegetable Growing in India*. Oxford and IBH Pub. Co., New Delhi, India.

Katyal, S.L., 1977. Celery. In: *Vegetable Growing In India*. Oxford and IBH Publication, New Delhi, India.

Khan, N.A., Haq, F., Bengum and Huseain, M.B., 1982. Studies on *Coriandrum sativum* Linn. I. Chemical investigation of the seed. *Bangladesh Sci. and Indust. Res.*, 17(3): 172–177.

Khanna, R.K., Sharma, O.S., Raina, R.M., Sinha, S. and Singh, A., 1985. The essential oil of *Origanum majorana* raised on saline alkaline soil. *Indian Perfumer*, 29(3–4): 171–176.

Kochhar, S.L., 1998. *Economic Botany in the Tropics.* Macmilan India Ltd., Delhi, India.

Krishan, S., Kamath, H.R., Kudva, K.T. and Kudva, K.G. Cinnamon leaf oil. *J. Scient. 2nd Res. India*, 4: 464–466.

Krishnamoorthy, B. and Rema, J., 1991. Allspice. *Spice India*, 4(10): 9–10.

Krishnamoorthy, B. and Rema, J., 1991. Cassia and cinnamon. *Spice India*, 4(3): 10–12.

Krishnamurthy, M. and Sampathu, S.R., 1988. Anti-oxidant properties of Kokum rind. *J. Fd. Sci. Technol.*, 25(1): 44–45.

Krishnamurthy, N., Lewis, Y.S. and Ravindranath, B., 1982. Chemical constituents of Kokum fruit rind. *J. Fd. Sci. Technol.*, 19(3): 97–100.

Kumar, P., 1990. The saffron story. *Indian Spices*, 27(1): 5–12.

Lawrence B.M., 1987. Laurel lead oil. *Perfume and Flavourist*, 12(8)-(9): 71–73.

Lawrence, B.M., 1979. Progress in essential oils: Fennel and peppermint oils. *Perfumer and Flavounist*, 4(5): 9–10, 12–13.

Lawrence, B.M., 1981. Progress in essential oils. *Perfumer and Flavourist*, 6(5): 27–34.

Lawrence, B.M., 1983 and 1984. Majoram oil. *Perfumer and Flavourist*, 8: 67, 9: 54–56.

Lawrence, B.M., 1984. Progress in essential oils: Cassia, Cl, Fennel. *Perfume and Reave*, 9(2): 23–30.

Lawrence, B.M., 1984. Progress in essential oils: Fennel and marjoram oils. *Pefumer and Flavourist*, 9(1): 43–60.

Lawrence, B.M., 1985. Progress in essential oils: Caraway. *Perfumer and Flavourist*, 10(1): 34–38, 51–53.

Lawrence, B.M., 1989. Progress in essential oils: Basil oil (A review). *Perfumer and Flavourist*, 14(5): 45–5.

Lawrence, B.M., 1989. Progress in essential oils: Celery seed and leaf oil (Review). *Perfumer and Flavourist*, 14(5): 52–53.

Lawrence, B.M., 1993. Essential Oils, 1988–91 (including Ajowan oil). Allured Pub. Corpn, Carol Streem, Illinois 60188, USA Vol. 4 pp. 226.

Lewis, Y.S. and Neelkanth, S., 1964. The chemistry, biochemistry and technology of tamarind. *J. Sci Res.*, 23(5): 204–206.

Lewis, Y.S., 1987. Tamarind pulp, paste, concentrate and powder. Status Paper *Souv. Nat. Symp. on Spice Industries*, (Ed.) J.S. Pruthi. AFST (I) Delhi, pp. 62–65.

Lowman, M.S., 1946. Savory herbs: Culture and use. *USDA Farmers Bull No. 1977*, pp. 33.

Maarse, H.O.S.F.H.L. Van, 1973. Volatile oil of *Origanum vulgare* L. spp. *vulgare* I. Qualitative composition of the oil. *Flavur Ind.*, 4(11): 447–481.

Marsh, A.C., Moss, M.K. and Murph, E.W., 1977. Composition of foods, spices and herbs: Raw Processed, prepared. Washington DC, USDA *Agri. Res. Serv. Handbook*, 8–2.

Mitra, Roma, 1998. Ethno-economic significance of the common myrtle: A plant sacred to Greeks and Romans. *Ethnobotany*, 10: 1–5.

Nagaraja, K.V., Manjunath, M.V., and Nalini, M.L., 1975. Chemical composition of commercial tamarind juice concentrate. *Indian Fd. Packer*, 29(5): 17–20.

Nair, M.K., 1978. Clove and nutmeg. *Indian Fmg.*, 28(4): 10–13, 35.

Nair, P.C.S. and Mathew, L., 1969. Vanilla. *Indian Spices*, 6(4): 2–4.

Narayanan, R. and Ars, Kartha, 1962. The glyceride structure of the seed and mace fats from M. Fragrans *J. Scient Ind. Res.*, 21: 442–444.

Nath, S.C., Hazarika, A.K. and Singh, R.S., 1994. Essential oils of leaves of *Cinnamonum tamala* Nees and Eberm. from North East India. *J. Spices and Aromatic Crops*, 3(1): 33–35.

Nauriyal, J.P., Gupta, R. and George, C.K., 1977. Saffron in India. *Arecanut and Spices Bull.*, 8 (3): 59–72.

Nigam, M.C. and Ahmed, A., 1990. The essential oil of *Cinnamonum tamala*. II. *Pafai Journal*, 12(2): 21–22.

Nigam, S.S. and Purohit, R.M., 1961. Chemical examination of the essential oil from the leaves of *Murrays koemigi* (Linn.) Spreng. (Indian Curry Leaf). *Perfum Essent. Oil Rec.* (London), 52: 153.

Nishioka, I., 1983. Seven aromatic compounds from bark of *C. cassia*. *Phytochemistry*, 22(1): 215.

Nohara T. *et al.*, 1982. Two novel diterpenes from the bark of *C. cassia*. *Phytochemistry* 21(8): 2130–2132.

Okamura, N., Haraguch, H., Hashimoto, K. and Yogi, A., 1994. Flavonoids in *Rosmarinus officinalis* leaves. *Phytochemistry*, 37(5): 1463–1467.

Pamcker, P.M.B., Rao, B.S. and Simonson, J.L., 1926. Chemical examination of *Kaempferia galangal* L. *J. Indian Instt. Sci.*, 9A: 133.

Passet, 1979. Chemical variability within thyme, it manifestation and its significance. *Perfume Cosmet. Aromes*, 28: 39–42.

Pathak S.P. and Ojha, V.N., 1957. The component glyceride of nutmeg mutter. *J. Sci. Fd. Agric.*, 8: 537–540.

Patil, D.A., 2008. *Medicinal Plants: History Culture and Usage*. Mangalam Publishers and Distributors, Delhi, India.

Patil, D.A., 2008. *Origin of Plant Names*. Daya Publishing House, New Delhi, India.

Patil, D.A., 2008. *Useful Plants: Origin History and Civilization*. Navyug Publishers and Distributors, Delhi, India.

Pino, J.A., Borges, P. and Mdlinedo, B., 1991. Essential oil of nutmeg: Extraction and chemical composition. *Reg. Agroquim Technol Aliment*, 31(3): 411–417.

Preiminger, V. *et al.*, 1965. Occurrence of alkaloids in op poppy seed (*Papaver sommiferum*). *J. Agri. Food Change*, 29: 1232–1235.

Prez Alogo, M.J. *et al.*, 1995. Composition of the essential oils of *O. basilicum* var. *glabratim* and *Rosemarines officintis* from Turke. *J. Essential Oil Res.*, 7(1.) 73–75.

Pruthi, J.S., 1958. Horse radish. In: *Spices and Condiments*, 5th edn. National Book Trust, New Delhi, India, pp. 138–140.

Pruthi, J.S., 1978. Savor. In: *Spices and Condiments,* 5th edn. National Book Trust, New Delhi, India, pp. 210–211.

Pruthi, J.S., 1980. Basil or sweet basil. In: *Spices and Condiments,* 5th edn. National Book Trust, New Delhi, India, pp. 29–35.

Pruthi, J.S., 1980. *Spice and Condiments: Chemistry, Microbiology, Technology.* Special Supp. IV. Advances in Food Research 12: 1–450.

Pruthi, J.S., 1980. Spices as food colourants. *J. Bev. Food World,* 7(4): 11.

Pruthi, J.S., 1982. Cinamon In: *Spices and Condiments,* 5th edn. National Book Trust, New Delhi, India, pp. 86–90, and Cassia *Ibid* pp. 69–80.

Pruthi, J.S., 1988. Long pepper. In: *Spices and Condiments,* 5th edn. National Book Trust, New Delhi, India, pp. 191–193.

Pruthi, J.S., 1998. Ajowan. In: *Spices and Condiments,* 5th edn. National Book Trust, New Delhi, India, pp. 5–8.

Pruthi, J.S., 1998. Allspice In: *Spices and Condiments,* 5th edn. National Book Trust, New Delhi, India.

Pruthi, J.S., 1998. Amchur. Star Anise. In: *Spices and Condiments,* 5th edn. National Book Trust, New Delhi, India, pp. 12–14.

Pruthi, J.S., 1998. Asafoetida. In: *Spices and Condiments,* 5th edn. National Book Trust, New Delhi, India, pp. 22–28.

Pruthi, J.S., 1998. Bay or laurel leaves. In: *Spices and Condiments,* 5th edn. National Book Trust, New Delhi, India.

Pruthi, J.S., 1998. Black cumin. In: *Spices and Condiments,* 5th edn. National Book Trust, New Delhi, India, pp. 112–115.

Pruthi, J.S., 1998. Clove. In: *Spices and Condiments,* 5th edn. National Book Trust, New Delhi, India, pp. 92–98.

Pruthi, J.S., 1998. Coriander In: *Spices and Condiments,* 5th edn. National Book Trust, New Delhi, India, pp. 99–103.

Pruthi, J.S., 1998. Cumin. In: *Spices and Condiments,* 5th edn. National Book Trust, New Delhi, India, pp. 104–108.

Pruthi, J.S., 1998. Dill and Indian dill. In: *Spices and Condiments,* 5th edn. National Book Trust, New Delhi, India, pp. 111–114.

Pruthi, J.S., 1998. Fenugreeek In: *Spices and Condiments,* 5th edn. National Book Trust, New Delhi, India, pp. 121 –124.

Pruthi, J.S., 1998. Galangal. In: *Spices and Condiments,* 5th edn. National Book Trust, New Delhi, India.

Pruthi, J.S., 1998. Garlic In: *Spices and Condiments,* 5th edn. National Book Trust, New Delhi, India, pp. 126–133.

Pruthi, J.S., 1998. Hussop. In: *Spices and Condiments,* 5th edn. National Book Trust, New Delhi, India, pp. 141–143.

Pruthi, J.S., 1998. Jumiper. In: *Spices and Condiments,* 5th edn. National Book Trust, New Delhi, India, pp. 143–147.

Pruthi, J.S., 1998. Kokum. In: *Spices and Condiments,* 5th edn. National Book Trust, New Delhi, India, pp. 147–149.

Pruthi, J.S., 1998. Leek. In: *Spices and Condiments,* 5th edn. National Book Trust, New Delhi, India, pp. 1–322.

Pruthi, J.S., 1998. Mace and nutmeg. In: *Spices and Condiments,* 5th edn. National Book Trust, New Delhi, India, pp. 151–155; Nutmeg, pp. 167–172.

Pruthi, J.S., 1998. Majoram. In: *Spices and Condiments,* 5th edn. National Book Trust, New Delhi, India, pp. 1–320.

Pruthi, J.S., 1998. Mint/jap, mint. In: *Spices and Condiments,* 5th edn. National Book Trust, New Delhi, India, pp. 158–162.

Pruthi, J.S., 1998. Mustard. In: *Spices and Condiments,* 5th edn. National Book Trust, New Delhi, India, pp. 164.

Pruthi, J.S., 1998. Onion. In: *Spices and Condiments,* 5th edn. National Book Trust, New Delhi, India, pp. 173–175.

Pruthi, J.S., 1998. Origanum. In: *Spices and Condiments,* 5th edn. National Book Trust, New Delhi, India, pp. 1–340.

Pruthi, J.S., 1998. Parsely. In: *Spices and Condiments,* 3rd edn. National Book Trust, New Delhi, India, pp. 177–180.

Pruthi, J.S., 1998. Peppermint. In: *Spices and Condiments,* 5th edn. National Book Trust, New Delhi, India.

Pruthi, J.S., 1998. Poppy seeds. In: *Spices and Condiments,* 5th edn. National Book Trust, New Delhi, India, pp. 194–198.

Pruthi, J.S., 1998. Rosemary. In: *Spices and Condiments*, 5th edn. National Book Trust, New Delhi, India.

Pruthi, J.S., 1998. Saffron. In: *Spices and Condiments*, 5th edn. National Book Trust, New Delhi, India, pp. 203–206.

Pruthi, J.S., 1998. Spearmint. In: *Spices and Condiments*, 5th edn. National Book Trust, New Delhi, India, pp. 211–213.

Pruthi, J.S., 1998. *Spices and Condiments*, 5th edn. National Book Trust, New Delhi, India, pp. 1–322.

Pruthi, J.S., 1998. Star Anise. In: *Spices and Condiments*, 5th edn. National Book Trust, New Delhi, India, pp. 215–217.

Pruthi, J.S., 1998. Sweet flage or calamns. In: *Spices and Condiments*, 5th edn. National Book Trust, New Delhi, India, pp. 217–219

Pruthi, J.S., 1998. Tamarind. In: *Spices and Condiments*, 5th edn. National Book Trust, New Delhi, India, pp. 217–219.

Pruthi, J.S., 1998. Tarragaon. In: *Spices and Condiments*, 5th edn. National Book Trust, New Delhi, India, pp. 222–223.

Pruthi, J.S., 1998. Thyme. In: *Spices and Condiments*, 5th edn. National Book Trust, New Delhi, India.

Pruthis, J.S., 1998. Celery. In: *Spices and Condiments*, 5th edn. National Book Trust, New Delhi, India, pp. 1–322.

Puhan, S.P.S., 1983. European dill (*Anethum graveolens*): A crop of Pharmaceutical value. *Indian Fd. Dig.*, 16(12): 21–22.

Puri, S.C., Vasisht, V.N. and Atal, C.K., 1968. The essential oil of Kulu coriander. *Indian Oil Soap J.*, p. 34.

Radhakrishnan, V.V., 1992. Cinnamon: The spice bark. *Spice India*, 5(4): 11–13.

Raghavan B. *et al.*, 1974. Asafoetida–II. Chemical composition and physio-chemical properties. *Flavour Ind.*, 5(7/8): 179–181.

Rajanikanth, B., Ravindranah, B. and Shandaranarayana, M.L., 1984. Volatile polyulphides of asafoetida. *Phytochem.*, 23(4): 899–900.

Ramachandraiah, D.S. *et al.*, 1984. Studies on Indian essential oils, deodar seed, davana, sweet majoram and pudina (mint). *Indian Perfumer*, 281(1): 10–16.

Ravid, U., Putievsky, E. and Shir, N., 1983. The volatile components of oleoresins and the essential oils of *Foemicullm vulgare* in Israel. *J. Prod.*, 46(6): 848–851.

Salgueiro, L.M.R., 1991. Essential oil of endemic Thymus species from Portugal. *Flavour and Fragrance Journal*, 7(3): 159–162

Sampathu S.R. and Krihnamurthy, N., 1982. Processing and utilization of Kokum. *Indian Coco, Arecanut and Spices J.*, 6(1): 12–13.

Sen, A.R., Sardar, P.K., Sil, S. and Mathew, T.V., 1973. Importance of volatile oil and cold water extract estimation in analysis of cumin (Jira). *J. Fd. Sci. Technol.*, 10(4): 187.

Senanyake, U.M., Lee, T.H. and Wills, R.B.H., 1981. Volatile constituents of cinnamon oils. *J. Agnia Fd. Chem.*, 26(4): 822–824.

Shakhova M.F., 1978. Vitamins of peppermint and the use Tr. Mostk *O-va Ispyt Prir*, 54: 62–63.

Shankarachara, N.B. and Natarajan, C.P., 1971. Coriander chemistry, technology and uses. *Indian Spices*, 8(2): 4–13.

Shankarachara, N.B. and Natrajan, C.P., 1971. Leaf spice: Chemical composition and uses. *Indian Fd. Packer*, 25: 28.

Shankaracharya, N.B. and Natarajan, C.P., 1971. Chemical composition and uses of cumin. *Indian Fd. Packer*, 25(6): 22–28.

Shankaracharya, N.B. and Natarajan, C.P., 1971. Leaf spices: Chemical composition and uses. *Indian Fd. Packer*, 25(2): 29–40.

Shankaracharya, N.B. and Natarajan, C.P., 1972. Fenugreek: Chemical composition and uses. *Indian Spices J.*, 2(1): 1–11.

Shankaracharya, N.B. and Natarajan, C.P., 1973. Vanilla: Chemistry, technology and uses. *Indian Fd. Packer*, 26(3): 29–36.

Shankarcharya, N.B., Anandaraman, S. and Natarajan, C.P., 1993. Chemical composition of coriander varieties and changes on roarting. *J. Plant Crops*, 1: 79.

Sharma, B.R. and Arora, S.I., 1991. Fenugreek. *Spice India's*, 4(4): 19–21.

Simmonds, N.W., 1976. *Evolution of Crop Plants*. Longman, London (U.K.) and New York (U.S.A.).

Simon, J.E. and Quinn, J., 1988. Characterization of essential oil of Parsle. *J. Agric Fd. Chem.*, 36: 467–472.

Singh, Bhanu Pratap, 1989. Tejpat: An Indian free spice. *Spice India*, 2(8): 5–6.

Singh, Gurmit *et al.*, 1959. Composition of *O. vulgare* oil from plant growing in J&K. *J. Scient. Ind. Res.*, 18B(3): 128–129.

Singh, H.P., Sivaraman, K. and Selvanc, M. Tamil, 2002. *Indian Spices: Production and Utilization.* Coconut Development Board, Ministry of Agriculture Government of India, Kochi, India, Publication No. 108.

Singh, H.P.K., Sivaraman and Selvan, Tamil, 2002. *Indian Spices: Production and Utilization.* Coconut Development Board, Ministry of Agriculture Government of India, Kochi, India, Publication No. 108.

Srinathan, R.P., Padha, C.D., Talwar, Y.P., Jamwal, R.K., Chopra, M.M. and Rao, P.K., 1979. Essential oils from the leaves of *Cinnamonum tamala* Nees and Eberm growing in Himachal Pradesh. *Indian Perfumer*, 23127: 75–78.

Srinivas, H. and Rao, M.S. Narasinya, 1981. Studies on the proteins of poppy seeds. *J. Agri. Food Chem.*, 29: 1232–1235.

Sriram, T.A., 1976. Tree spices: Retrospect and research needs. *Arecanut and Spices Bull.*, 8(4): 97–100.

Sriram, T.A., 1976. Tree spices: Retrospect and research needs. *Arecanut and Spices Bull.*, 8(4): 97–100.

Srivastav, V.K., Hill, D.C. and Skinger, S.J., 1976. Companions of some chemical characteristics of India: An Canadian Brassice seeds. *Indian J. Nutr. Dietet.*, 13(10): 336–342.

Srivastava J.G., 1971. Botanical identify of Vacha (Bachh) of the ayurvedic literature. *Q. J. Crude Drug Res.*, 11(2): 1734–1742.

Subrhmanyam, V. and Srinivasan, M., 1955. Asafoetida: Its origin, nature and place in human dietary and medicine. *Bull. CFTRI*, 5(2): 27–29

Sundaravalli, D.E. and Shurpalekr, Kantha, 1981. Nutritional evaluation of Kokam (*Garaima indica*) and Mango (*Mangifera indica*) fat. *J. Technol Ass. India*, 13(3): 116–119.

Tenber, H. and Hermann, K., 1978. Flaverol glycosides of leaves and fruits of dill (*Anethum graveolens* L.J.) It flavour phenols 2. Lebensmittelunters. U. *Forsch*, 167(2): 101–104.

Tucker, A.O. and Maciarello, M.J., 1994. Overganobotany, chemistry and cultivation. In: *Spices, Herbs and Edible Fungi*, (Ed.) G. Charalambous. Elsevier Science Publishers, New York. USA, pp. 439–457.

Tucker, A.O., Machiarell, M.J. and Howell, J.T., 1980 Botanical aspects of commercial sage. *Econ. Bot.*, 34(1): 16–19.

Varrier, P.K., 2001. *Indian Medicinal Plants Vols. 1–5* (Repr. Ed.). Orient Longman Ltd., Hyderabad, India.

Varshne, S.C., 1992. Essential oil industry in India: Growth, potential and constraints. *Indian Perfumer*, 36(3): 28–231.

Veek, M.E. and Russell, G.F., 1973. Chemical and sensory properties of Pimento leaf oil. *J. Fd. Sci.*, 38(6): 1028–1031.

Venkatachalam, K.V., Kjonas, R. and Croteau, R., 1984. Development and essential oil content of secretory glands of sage. *Pl. Physiol.*, 76(1): 148–150.

Virmani, O.P., Singh, Pratap and Hussain, Akhtar, 1986. Current Status of medicinal plants industry in India. *Indian Drugs*, 17(10): 318–340.

Wijesekera, R.O.B., 1979. The chemistry and technology of Cinnamon. *CRC Critical Reviews in Sci. Nutr.*, 10(1): 1–30.

Wijesekera, R.O.B., *et al.*, 1975. Essential oils IV Recent studies on the volatile oils of cinnamon. *J. Nat. Sci. Council, Sri Lanka*, 2011: 45.

Wijesekera, R.O.B., Jayawardene, A.I. and Oadse, L.S. Raja, 1974. Volatile constituents of leaf, stem, root oils of cinnamon. *J. Sci. Fd. Agric.*, 25: 1211–1220.

Wu, T.S. Leu, L., Chan, Y.Y., Yu, S.M., Teng, C.M. and Su, J.D., 1994. Lignaus and an aromatic acid from *Cinnamonum* Philippines. *Phytochemintry*, 36(3): 785–789.

Yawalkar, K.S., 1969. *Vegetable Crops of India*. Agri. Hort. Publ. House, Nagpur, pp. 1–300.

Index